Advances in Intelligent Systems and Computing

Volume 221

Series Editor

J. Kacprzyk, Warsaw, Poland

For further volumes:
http://www.springer.com/series/11156

Javier Bajo Pérez · Juan Manuel Corchado Rodríguez
Johannes Fähndrich · Philippe Mathieu
Andrew Campbell
María del Carmen Suárez-Figueroa
Emmanuel Adam · Antonio Fernández-Caballero
Ramon Hermoso · María N. Moreno
Editors

Trends in Practical Applications of Agents and Multiagent Systems

11th International Conference on Practical Applications of Agents and Multi-Agent Systems

Editors
Javier Bajo Pérez
Departamento de Inteligencia Artificial
Facultad de Informática
Universidad Politécnica de Madrid
Madrid, España

Juan Manuel Corchado Rodríguez
Departamento de Informática y Automática
Facultad de Ciencias
Universidad de Salamanca
Salamanca, Spain

Johannes Fähndrich
Technische Universität Berlin
Faculty of Electrical Engineering and Computer Science
Berlin, Germany

Philippe Mathieu
Université des Sciences et Technologies de Lille
LIFL (Laboratoire d'Informatique Fondamentale de Lille)
Villeneuve d'Ascq Cedex, France

Andrew Campbell
Department of Computer Science
Dartmouth College
Hanover
USA

María del Carmen Suárez-Figueroa
Departamento de Inteligencia Artificial
Facultad de Informática
Universidad Politécnica de Madrid
Madrid, España

Emmanuel Adam
LAMIH (UMR CNRS 8530)
Universite de Valenciennes
Valenciennes - Cedex 9, France

Antonio Fernández-Caballero
Departamento de Informática
University of Castilla-La Mancha
Campus Universitario s/n
Albacete, Spain

Ramon Hermoso
Universidad Rey Juan Carlos
Móstoles, Madrid
Spain

María N. Moreno
Departamento de Informática y Automática
Facultad de Ciencias
Universidad de Salamanca
Salamanca, Spain

ISSN 2194-5357 ISSN 2194-5365 (electronic)
ISBN 978-3-319-00562-1 ISBN 978-3-319-00563-8 (eBook)
DOI 10.1007/978-3-319-00563-8
Springer Cham Heidelberg New York Dordrecht London

Library of Congress Control Number: 2013937540

Printed on acid-free paper

Springer is part of Springer Science+Business Media (www.springer.com)

Preface

PAAMS'13 Special Sessions are a very useful tool in order to complement the regular program with new or emerging topics of particular interest to the participating community. Special Sessions that emphasized on multi-disciplinary and transversal aspects, as well as cutting-edge topics were especially encouraged and welcome.

Research on Agents and Multi-Agent Systems has matured during the last decade and many effective applications of this technology are now deployed. An international forum to present and discuss the latest scientific developments and their effective applications, to assess the impact of the approach, and to facilitate technology transfer, has become a necessity.

PAAMS, the International Conference on Practical Applications of Agents and Multi-Agent Systems is an evolution of the International Workshop on Practical Applications of Agents and Multi-Agent Systems. PAAMS is an international yearly tribune to present, to discuss, and to disseminate the latest developments and the most important outcomes related to real-world applications. It provides a unique opportunity to bring multi-disciplinary experts, academics and practitioners together to exchange their experience in the development of Agents and Multi-Agent Systems.

This volume presents the papers that have been accepted for the 2013 edition in the special sessions: Assessing Agent Applications & Self-Explaining Agents, Agents Behaviours and Artificial Markets, Agents and Mobility, Intelligent components producing and consuming knowledge and data, COREMAS: COoperative and RE-configurable MultiAgent System for Industrial Environments, Multi-Agent Systems for Safety and Security, TINMAS: Trust, Incentives and Norms in open Multi-Agent Systems, WebMiRes: Web Mining and Recommender systems.

We would like to thank all the contributing authors, as well as the members of the Program Committees of the Special Sessions and the Organizing Committee for their hard and highly valuable work. Their work has helped to contribute to the success of the PAAMS'13 event. Thanks for your help, PAAMS'13 wouldn't exist without your contribution.

We thank the sponsors (IEEE Systems Man and Cybernetics Society Spain, AEPIA Asociación Española para la Inteligencia Artificial, AFIA French Association for Artificial Intelligence, CNRS Centre national de la recherche scientifique), the Local Organization members and the Program Committee members for their hard work, which was essential for the success of PAAMS'13.

Juan Manuel Corchado Rodríguez
Javier Bajo Pérez
PAAMS'13 Organizing Co-chairs

Organization

Special Sessions

Assessing Agent Applications & Self-Explaining Agents
Agents Behaviours and Artificial Markets
Agents and Mobility
Intelligent components producing and consuming knowledge and data
COREMAS: COoperative and RE-configurable MultiAgent System for Industrial Environments.
Multi-Agent Systems for Safety and Security
TINMAS: Trust, Incentives and Norms in open Multi-Agent Systems.
WebMiRes: Web Mining and Recommender systems.

Special Session on Assessing Agent Applications & Self-Explaining Agents

Sebastian Ahrndt (Chair)	DAI-Labor, Technische Universität Berlin, Germany
Johannes Fähndrich (Chair)	DAI-Labor, Technische Universität Berlin, Germany
Benjamin Hirsch (Chair)	EBTIC / Khalifa University, United Arab Emirates
Tina Balke (Chair)	University of Bayreuth, Germany
Marco Lützenberger (Chair)	DAI-Labor, Technical University of Berlin, Germany
Dawud Gordon	TECO, Karlsruhe Institut of Technology, Germany
Christian Müller-Schloer	Leibniz University Hannover, Germany

Special Session on Agents Behaviours and Artificial Markets

Philippe Mathieu (chair)	University Lille, France
Dr. Bruno Beaufils	University of Lille, France
Pr. Olivier Brandouy	University of Paris, France
Pr. Andrea Consiglio	University of Palermo, Italy
Dr. Florian Hauser	University of Innsbruck, Austria
Pr. Philippe Mathieu	University of Lille, France
Pr. Juan Pavón	University Complutense Madrid, Spain
Dr. Marco Raberto	University of Genoa, Italy
Dr. Roger Waldeck	Telecom Bretagne, UK
Dr. Hector Zenil	University of Sheffield, UK

Special Session on Agents and Mobility

Andrew Campbell	Darthmouth College, USA
Javier Bajo	Polytechnic University of Madrid, Spain
Gabriel Villarrubia	University of Salamanca, Spain
Juan F. De Paz	University of Salamanca, Spain
Fernando de la Prieta	University of Salamanca, Spain

Special Session on Intelligent Components Producing and Consuming Knowledge and Data

Mari Carmen Suárez-Figueroa (chair)	Ontology Engineering Group (OEG), Universidad Politécnica de Madrid, Spain
Valentina Tamma (chair)	University of Liverpool, United Kingdom
Stefan Decker	Digital Enterprise Research Institute, Ireland
Mariano Fernández-López	Universidad San Pablo CEU, Spain
Chiara Ghidini	Fondazione Bruno Kessler, FBK, Italy
Massimo Paolucci	DOCOMO Euro-Labs, Germany
Terry Payne	University of Liverpool, United Kingdom
Carlos Pedrinaci	Open University, United Kingdom
Murat Sensoy	University of Aberdeen, United Kingdom
Luciano Serafini	Fondazione Bruno Kessler, FBK, Italy
Serena Villata	INRIA Sophia Antipolis, France
Sonia Bergamaschi	Universita' di Modena e Reggio Emilia, Italy
Boris Villazón	Terrazas, iSOCO, Spain

Special Session COREMAS: COoperative and RE-configurable MultiAgent System for Industial Environments

Emmanuel Adam	University of Valenciennes, France
Paulo Leitao	Polytechnic Institute of Bragança, Portugal

Special Session on Multi-Agent Systems for Safety and Security

Antonio Fernández-Caballero	University of Castilla-La Mancha, Spain
Elena María Navarro Martínez	University of Castilla-La Mancha, Spain

Special Session on Trust, Incentives and Norms in Open Multi-Agent Systems

Ana Paula Rocha (Chair)	University of Porto, Portugal
Henrique Lopes Cardoso (Chair)	University of Porto, Portugal
Olivier Boissier (Chair)	ENSM Saint-Etienne, France
Ramón Hermoso (Chair)	Universidad Rey Juan Carlos, Spain
Cristiano Castelfranchi	ISTC-CNR, Italy
Daniel Villatoro	IIIA-CSIC, Spain
Eugenio Oliveira	University of Porto, Portugal
Joana Urbano	University of Porto, Portugal
Jordi Sabater-Mir	IIIA-CSIC, Spain
Laurent Vercouter	University of Rouen, France
Nicoletta Fornara	University of Lugano, Switzerland
Patrice Caire	University of Luxemburg, Luxemburg
Roberto Centeno	UNED, Spain
Sascha Ossowski	Universidad Rey Juan Carlos, Spain
Wamberto Vasconcelos	University of Aberdeen, Scotland
Waqar ul Qounain	University of The Punjab, Pakistan

Special Session WebMiRes: Web Mining and Recommender systems

María Moreno (chair)	University of Salamanca, Spain
Longbing Cao (chair)	University of Technology, Australia
Ana María Almeida	Institute of Engineering of Porto (Portugal)
James Bailey	University of Melbourne, Australia
Yolanda Blanco Fernández	University of Vigo, Spain
Sanjay Chawla	University of Sydney, Australia
Peter Christen	Australian National University
Rafael Corchuelo	University of Sevilla, Spain
Chris Cornelis	Ghent University, Belgium
María José del Jesús Díaz	University of Jaén, Spain
Anne Laurent	University of Montpellier 2, France
Vivian López Batista	University of Salamanca, Spain
Joel Pinho Lucas	Mobjoy, Brazil
Constantino Martins	Institute of Engineering of Porto, Portugal
Simeon Simoff	University of Western Sydney, Australia
Haixun Wang	Microsoft Resarch Asia
Guandong Xu	Victoria University, Australia
Yanchun Zhang	Victoria University, Australia

Organizing Committee

Juan M. Corchado (Chairman)	University of Salamanca, Spain
Javier Bajo (Co-Chairman)	Pontifical University of Salamanca, Spain
Gabriel Villarrubia	University of Salamanca, Spain
Alejandro Sánchez	University of Salamanca, Spain
Juan F. De Paz	University of Salamanca, Spain
Sara Rodríguez	University of Salamanca, Spain
Dante I. Tapia	University of Salamanca, Spain
Fernando de la Prieta Pintado	University of Salamanca, Spain
Davinia Carolina Zato Domínguez	University of Salamanca, Spain
Elena García	University of Salamanca, Spain
Roberto González	University of Salamanca, Spain

Reviewers

Philippe Mathieu	University of Lille1, France
Bruno Beaufils	University of Lille1, France
Hector Zenil	University of Sheffield, UK
Juan Pavón Mestra	Universidad Complutense Madrid, Spain
Olivier Brandouy	University of Paris 1, France
Marco Raberto	University of Genoa, Italy
Florian Hauser	*University* of Innsbruck, Austria
Roger Waldeck	Telecom Bretagne Brest, France
Andrea Consiglio	Università degli Studi di Palermo, Italy

Contents

Special Session on Intelligent Components Producing and Consuming Knowledge and Data (ICP)

Special Sessions COoperative and RE-configurable MultiAgent System (COREMAS)

Special Session on Multi-Agent Systems for Multi-sensor Activity Interpretation (MASMAI)

A Stylized Software Model to Explore the Free Market Equality/Efficiency Tradeoff

Hugues Bersini and Nicolas van Zeebroeck

Abstract. This paper provides an agent-based software exploration of the well-known free market efficiency/equality trade-off. Our study simulates the interaction of agents producing, trading and consuming goods within different market structures, and looks at how efficient the producers/consumers mapping turn out to be as well as the resulting distribution of welfare among agents at the end of an arbitrarily large number of iterations. A competitive market is compared with a random one. Our results confirm that the superior efficiency of the competitive market (an effective producers/consumers mapping and a superior aggregative welfare) comes at a very high price in terms of inequality (above all when severe budget constraints are in play).

Keywords: ABM, free market, equality/efficiency trade-off.

1 Introduction

A classical disputed question regarding the effect of free market economy on the social welfare is the right balance between equality and efficiency called by Okun [1]: the big tradeoff. Part of the problem lies in the difficulty to appropriately define these two notions. The eternal question of equality, famously debated and popularized by, among the most modern thinkers, Rawls, Dworkin, Sen, depends upon 1) the right currency for equality (primary goods, consumers utility, opportunity, capability,) and 2) the right distribution of this currency (pure equality, some form of minmax principles i.e. favoring at a given time a distribution that is to the greatest benefit of the least-advantaged agent or others). On the other hand, the question of

Hugues Bersini · Nicolas van Zeebroeck
IRIDIA - Universite Libre de Bruxelles - CP 194/6
ECARES - Solvay Brussels School of Economics and Management - CP 114
50, av. Franklin Roosevelt, Bruxelles 1050 - Belgium
e-mail: {nospambersini,nicolas.van.zeebroeck}@ulb.ac.be

J.B. Pérez et al. (Eds.): *Trends in Prac. Appl. of Agents & Multiagent Syst.*, AISC 221, pp. 1–8.
DOI: 10.1007/978-3-319-00563-8_1

economic efficiency is even more ambiguous. It was originally framed around the Pareto optimality for which no one well-being should be raised without as a consequence reducing someone else well-being. Many Pareto optima can be obtained on an imaginary axis, going from a pure utilitarian aggregative end (at which what really counts is to maximize the collective well-being) to a more equalitarian end (where what really counts is to maximize the well-being of the worst agent). Indeed Pareto optimum per se is completely unconcerned with the appropriate distribution of the economical profit. It is enough that the agents welfare simply grows in time as a result of the economical interactions, leaving completely unresolved the comparison of economic systems that either promote aggregate welfare, perfect equality or the improvement of the poorer to the expense of the richer.

Another classical definition of efficiency related with multi-agents competitive system is the allocative one, in which the system must guarantee that a resource is being produced by the most skillful producer and goes to someone who draws the greater utility out of it. Not surprisingly, although efficient according to this definition, such a competitive system, likely to promote the best producers and to feed the greediest consumers, may have little chance to equally distribute wealth.

Beyond this historical debate about which economical system (free or regulated) has to be privileged between an aggregative or a distributive one, there is another key efficiency criteria which is often left out of the discussion, originally due to Hayek pioneering insights: his metaphor of the market as a system of telecommunication. Market prices are primarily a means of collating and conveying information for the producers to adequately response to the consumers needs. Thus, though a very high price prevents most of the consumers to acquire a product, it is, in the same time, a very reliable information addressed to the producer that many consumers are desperately in need of such a product. It might well be possible that a distributive economy, flattening the prices and rendering most of the products affordable to all, and although morally very defendable, turns out to corrupt this distributed information transmission mechanism and make all economical agents to see their situation finally degrade in time. In the rest of the paper, we will designate such incapacity of the market to effectively map producers onto consumers as market failures (MF).

In order to address these different issues, a software stylized model is proposed comparing two very different structures of market that potentially should drive the collective welfare to the two extremes: aggregative on one side and distributive on the other. These two structures are first a double auction competitive market (in which buyers and sellers compete to outbid each other) and a random market (in which the matching between buyers and sellers is done in a purely random way). Following the description of the model, many experimental outcomes of many robust runs will be presented along three key dimensions: the Gini indices (regarding equality), the aggregate utility and the probability of market failures (both regarding efficiency).

2 The Model

The model maps onto a C# object oriented software. The main encompassing class, the World, contains one Market, either competitive or random, where a given number of agents have the opportunity to successively produce, sell, buy and consume. This world evolves through discrete ticks. At every tick, a randomly selected agent is given a chance to produce one unit of one product among n possible ones. In the absence of financial means (producing cost money and this money leaks out of the system, all other processes leading to money transfers between agents), another random agent is selected until the production occurs. The market then attempts to execute one transaction that involves one buyer and one seller marketing one unit of a given product. If no transaction turns out to be possible, on account of an impossible pairing between buyers and sellers, the model raises a market failure (equivalent to an exception in the C# program). Once acquired by the buyer, the product is immediately consumed during the same tick and converted into utility according to his associated taste. Every agent starts with the same amount of money at the beginning of the simulation (allowing him to produce goods). Agents are distinctively characterized by two crucial factors which are their skills (influencing their producing behavior, production prices amount to the skills) and their tastes (imprinting their consuming behavior, utility increase amount to the tastes). While individual skills and tastes, taken randomly between 0 and 1, vary among agent, the initial total amount of skills and tastes are normalized to 1. This is the departing point of agents differentiation during the simulation and the only initial cause for any further inequality growing among the agents. Both producer and consumer behaviors are strictly similar in the competitive and the random markets whereas seller and buyer behaviors are fundamentally different.

Once randomly selected, the producer first has to decide which product to make. Two factors influence his decision: his skills and the average price of the last m transactions. Knowing his skills to produce each product unit and the average price in the market (memorized during the m previous ticks), it is obvious to compute his expected profit for each product. After x productions of the same product, an agent can further specialize himself making the production cost randomly diminishing within a moving range. Skills are then renormalized to 1 with all other skills proportionally rising up. Once an agent buys a product, it is immediately consumed, with effect to increase the agent utility by the value of his taste for this product. Two versions of the simulation are considered. In the first one, the utility does not decrease with the consumption and the preference of the agents keeps constant in time. In such a case, a competitive simulation, just based on the expression of the utility, should ease the demarcation of the agents along the simulations. In a second version, and in line with basic microeconomics concept of diminishing marginal utility, the taste associated to the product just consumed decreases for the next consumption. All tastes are then renormalized to 1 with all other products taste rising up accordingly.

The competitive market is akin to a continuous double auction market in which agents bid to buy and sell products units. During a succession of steps, the market repeatedly invites two randomly selected agents to place asks and bids on one product they want to sell or purchase. At the first tick, the market is initialized with best-buying and best-selling offers for all the products on the market (bids at price null and asks at price max). Then a random seller is selected to place an ask for the most profitable product he has in stock (the proposed price should be below the best-selling offer and incurring the least expense (i.e. selecting the product with the highest skill), this price is finally set between the producers skill and the current best-selling offer). The market then looks whether this ask crosses the current best-buying offer on that particular product. If so, the transaction occurs, if not, the ask becomes the best-selling offer and the market turns to the buying part. The randomly selected competitive buyer shows the very symmetrical behavior. He first selects the most desirable product (one with the highest taste above the best-buying offer) and places a bid limited by his reservation price (the proposed price is set between the best-buying offer and the reservation price). The market looks whether this bid crosses the current best-selling offer. Once two offers cross, the transaction price is fixed as the buying offer price. If following a determined number of trials, no transaction is to be found, a market failure is reported.

The random market is much simpler, since the sellers and the buyers behave without particular interest. In this version, a random seller places an ask on a random product, on which a random buyer is invited to react. If the buyer reservation price is higher than the price asked by the seller, a transaction takes place, the price being randomly set between the two offers. Here again, if following a determined number of trials, no transaction turns possible, a market failure is reported.

Finally, in order to impose a budgetary constraint on the buyers behavior, the reservation price for any product is fixed as the taste multiplied by the current money endowed by the agent multiplied by a time index (the agent portion of the budget he wills to engage at every tick). Of course, in all cases, bids and asks are only posted if the agent has, respectively, enough money to cover it or has a unit of the product in stock (as a result of previous productions). Whatever initial conditions being set: number of agents, number of products, vector of tastes and skills for every agent, initial endowment of money for all agents, they are obviously exactly equal for both market simulations, the objective being to compare the competitive version of the market (supposedly more efficient) with the random one (supposedly more equalitarian).

3 The Results

Four key metrics can be measured out of the different simulations: utility (increasing by consumption), money (leaking out by production and then fluctuating according to the transactions), added value (the difference between the price earned by the seller and the production cost) and market failures. For the first three, the aggregate value over all agents is used as an indicator of the market efficiency while the Gini coefficient (computed again for all three) testifies of how unequal this market turns out to be. The market failures (labeled MF in the following) is also used as an indicator of the market efficiency, but in the sense originally given by Hayek. Our simulations are always executed in the presence of 50 agents, 10 products and during 50000 simulation steps.For the first set of simulations, each agent is endowed with 500 units of money (so no budgetary constraint is imposed at all) and the number of past transactions kept in memory to inform the producer on the most valuable products is 1000. Additionally the consumers do not see their taste decreasing in time as an outcome of their consumption. Typical and quite robust experimental results follow, first for the random market then the competitive one.

Random Market: Total Utility: 5390, Total Money: 24312, Gini Utility: 0.04, Gini Money: 0.007, MF: 0

Competitive Market: Total Utility: 9755, Total Money: 24491, Gini Utility: 0.27, Gini Money: 0.08, MF:0

The competitive market turns out to be much more efficient in aggregative terms but this superior efficiency comes at a very high price in terms of inequality, compared with a random market (the utility Gini index is seven times greater as a result of the competition). Distortions in utility and money tend to grow over time. The competitive market favors those with skill in demand and those with taste skillfully satisfied. If this difference in taste can be continuously expressed over the simulation, a self-amplifying pairing happens between the greedy consumers and their dedicate competent producers. In the case of a marginally decreasing consuming utility, results become quite different, now making the competitive and the random markets rather comparable.

Random Market: Total Utility: 5152, Total Money: 24244, Gini Utility: 0.02, Gini Money: 0.007, MF: 0

Competitive Market: Total Utility: 5424, Total Money: 24488, Gini Utility: 0.042, Gini Money: 0.004, MF:0

In the absence of any budgetary constraint and if the same tastes cannot be differentially expressed all over the simulation (since being alternatively up and down as a result of the consumption), the competitive and random markets turn out to be very equivalent both in terms of efficiency and equality.

For the remaining of the simulations and in agreement with classical economics, the agents will see their taste decreasing in time as an outcome of their consumption.

The next aspect that deserves a dedicate treatment is the impact of information on the competitive market, evaluated by gradually varying the number of past transactions taken into account during the production process (fixed to 1000 so far) i.e. the quality and the reliability of the information available to the producers to guide their productions towards the real consumers needs. Many simulations have been run where the producers exploit an increasing number of past transactions: 0, 1, 5, 10, 50, 100, 500, 1000, 5000, 10000, 50000. In the previous simulations discussed so far, this number has been settled to 1000. We compute the average aggregate utility as a function of this number and, surprisingly, the resulting curve is not monotonous. Below 100 past transactions available to the producers, the resulting competitive markets show an important number of failures with a pick at 10, demonstrating, unexpectedly, that a total ignorance of the past is even better than a very little knowledge. An increasing amount of information first dilutes the effective signal upon which producers base their decisions. Producers in those cases may be better off only focusing on their own costs than on their expected profits. We finally can observe the relevance of sufficient information for the competitive market to efficiently allocate the available resources (and 1000 past transactions seem to be an appropriate minimal threshold above which no improvement is observed).The last aspect of the model to be explored is the influence of the budgetary constraint on the behavior of the market. While maintaining all other features constant (50 agents, 10 products, information based on 1000 past transactions), the initial money endowment is being decreased: 100, 80, 60, 50, 25, 20, 15, 10. After showing many difficulties in running until the end of the simulation, the random version of the market simply stops executing at around an initial endowment of 25. Many agents go bankrupt and the simulation is being constantly interrupted by market failures. Both facts once again testify of the

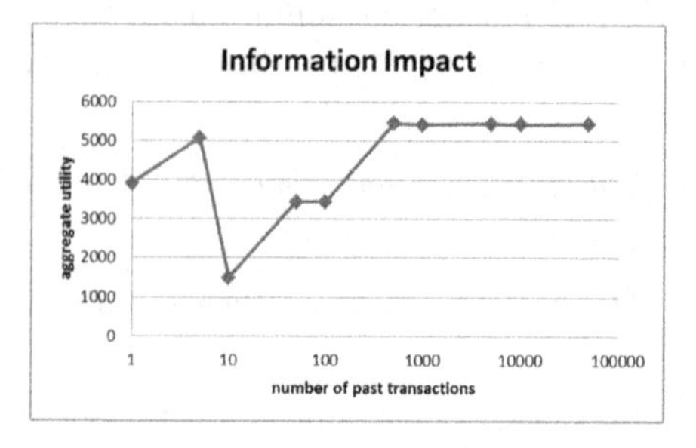

Fig. 1 Effect of the number of past transactions taken into account to optimize the production on the aggregate utility of the market

inefficiency of the random market to map the producers onto the consumers. The producers waste their money making products that the consumers are definitely not interested in.

As regards the competitive version, the table below indicates how the budget constraint really impacts the model as the initial endowment decreases. Although an initial budget of 20, 15 or 10, entails a few intermittent market failures, the model can now always keep running over the 50000 simulation ticks. The most striking fact of this table is the evolution of the utility Gini index as well as the added value one (for instance they respectively reach a pick of 0.25 and 0.20 for an initial budget of 10 by agent) that clearly shows a growing inequality as the money becomes scarcer. Again the market keeps being efficient but now to the large expense of equality. The competitive regime becomes much more selective towards the most skillful producers, the only ones who are effectively able to compete in the market. Budget constraint and money scarcity decrease the potential gains for producers but above all redirect them towards the best producers. Moreover, specialization acts as given the best producers even more marketing power. Budget constraints make the competition so severe that the smallest difference in skills is identified and reinforced. Figure 3 interestingly shows the correlation between the added value of the producer and his final utility as a consumer (i.e. established over all 50 agents). A clear positive correlation is observable between the added

Money	100	80	60	50	25	20	15	10
Utility	5465	5340	5464	5452	5433	5401	5330	5246
Money	4182	3275	2093	1735	494	284	83	17
Ad. Val.	3680	3720	3264	3380	2675	2400	1817	1424
G(Util)	0.05	0.061	0.077	0.070	0.088	0.12	0.16	0.25
G(Mon)	0.006	0.007	0.013	0.010	0.016	0.017	0.014	0.08
G(AV)	0.08	0.092	0.11	0.10	0.13	0.15	0.17	0.20
MF	0	0	0	0	0	26	270	914

Fig. 2 Summary of results (aggregate and Gini) obtained by gradually decreasing the initial budget possessed by every agent

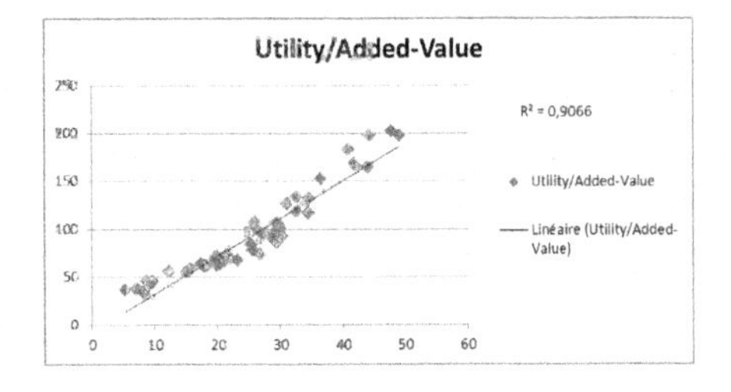

Fig. 3 Correlation between the added value of the producer and his final utility as a consumer (established over all 50 agents)

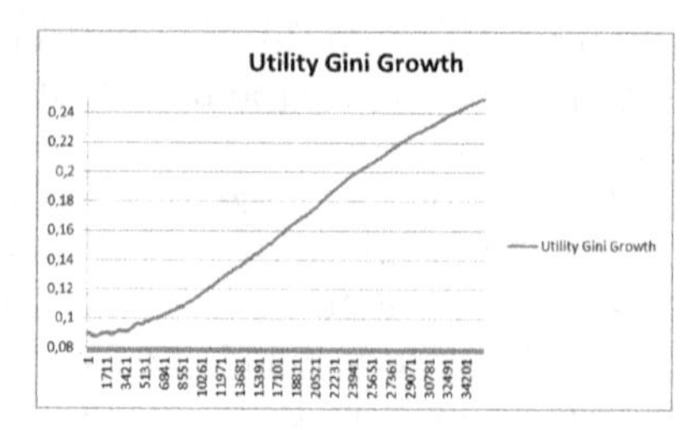

Fig. 4 Evolution in time of the Utility Gini Index for an initial budget of 10

value of the producer and his consumption (90% of the utility distribution is explained by the added value distribution). The greediest consumers turn out to be the best producers. As observable in fig.4, showing the evolution of the utility Gini index, inequality among the agents is on a fast growing trend. Competitive market acts in self-reinforcing the market dominance of producers who can benefit from the tiniest initial comparative advantage.

4 Conclusions

This paper describes a stylized simulation exercise in which we compare a double auction, quite aggressive, competitive market with a pure theoretical abstraction that represents a market in which producer and consumer matching is purely made on a random basis (under a natural set of constraints: budget constraint, no sale at loss rule for the producers)). Our main simulation results confirm the higher efficiency generally attributed to competitive markets first to simply map the consumers onto the producers then in maximizing the aggregative welfare. However in most of the studies of competitive markets, very little attention is paid to the equality in welfare distribution. Our results equally show this inequality explosion, above all in the case of budgetary constraints, when only the best producers can survive, make money and consume. Interestingly enough, at the starting of our simulation, all agents can be considered as equally ready and gifted to take part in the market, but its inherent competitive structure (in contrast with the random one) make an even negligible difference in skills to be greatly amplified with time. In line with most of the ethical philosophers, we can easily argue about the immoral nature of such an inequality amplifier mechanism (even when equality of opportunity is fully guaranteed) and the definitive need for a complementary equalizing system.

Reference

1. Okun, A.K.: Equality and efficiency, the big tradeoff. The Brookings Institution (1975)

Towards a Dynamic Negotiation Mechanism for QoS-Aware Service Markets

Claudia Di Napoli, Paolo Pisa, and Silvia Rossi

Abstract. The market value of commercial Service-Based Applications (SBAs) will depend not only on their functionality, but also on the value of QoS parameters referring to its not functional properties. These parameters are not static properties since they may vary according to the provision strategies of providers as well as the demand of users having their own preferences on the application's QoS values. In this paper we propose a negotiation-based mechanism among service providers and a user requesting a QoS-aware SBA to select services with suitable QoS values, i.e. values that once aggregated satisfy the user's requirements. The proposed mechanism simulates a market-based provision mechanism that allows to take into account the variability of service QoS attribute values typical of the future market of services, as well as to dynamically set the length of the negotiation process that is usually very time consuming.

Keywords: Market-based negotiation, Quality of Service, Service Selection.

1 Introduction

Service Based Applications (SBAs) are composed of autonomous and independent services each one provided with Quality of Service (QoS) attributes that take account of its non-functional properties (NFPs) such as cost, execution time, reliability, reputation, and so on [7]. In the future market of services QoS-aware SBAs

Claudia Di Napoli
Istituto di Cibernetica "E. Caianiello" - C.N.R., Via Campi Flegrei 34, 80078 Pozzuoli, Naples, Italy
e-mail: c.dinapoli@cib.na.cnr.it

Paolo Pisa · Silvia Rossi
Dipartimento di Ingegneria Elettrica e Tecnologie dell'Informazione, University of Naples "Federico II", Napoli, Italy
e-mail: silvia.rossi@unina.it

J.B. Pérez et al. (Eds.): *Trends in Prac. Appl. of Agents & Multiagent Syst.*, AISC 221, pp. 9–16.
DOI: 10.1007/978-3-319-00563-8_2

will be required by users that have their own preferences over the values of these attributes that may change in time according to dynamic circumstances. In order for QoS-aware SBAs to be delivered, the attribute values of their component services, once aggregated, have to meet the user requirements. In this context, it becomes crucial to provide service-oriented infrastructures with mechanisms enabling the selection of services with suitable QoS attribute values allowing to manage the dynamic nature of both QoS values, and QoS requirements.

In this paper we propose a negotiation-based mechanism among service providers and a service consumer to select the suitable services to compose QoS-aware SBAs through a market-based provision mechanism. The use of a negotiation-based mechanism allows to take into account the variability of service QoS attribute values typical of the future market of services since service providers may change these values during the negotiation according to their own provision strategies. For mandatory (i.e. not negotiable) QoS values the selection is obtained by solving a constraint satisfaction problem. Since negotiation can be computationally expensive, a set of experimental results were carried out to determine the parameters affecting the negotiation process to extract useful information to drive service consumers decisions about whether to proceed with the negotiation, under specific conditions, or not.

2 Selecting Services through Automated Negotiation

In this work it is assumed that a request for an SBA is expressed by a directed acyclic graph, called an *Abstract Workflow* (AW), specifying the functionality of each service component, and their functional dependence constraints, together with a quality attribute value representing the *QoS* required by the user for the application. AW nodes represent the required functionalities, referred to as *Abstract Services* (ASs), and AW arcs represent control and data dependencies among nodes. For each AS, a set of *Concrete Services* (CS) may be available on the market, each one provided by a specific *Service Provider* (SP) with QoS attributes whose values are set by the corresponding SP dynamically according to its market provision strategies. The user request is managed by a *Service Compositor* (SC), responsible for the selection of the CSs whose attribute values, once aggregated, satisfy the QoS required by the user. Both SPs and SC are represented by software agents able to negotiate.

The selection process is modelled as a negotiation process over the service quality attributes occurring among the SC and the SPs, available to provide the required services, that populate the multi-agent system. SPs issue their offers to the SC by specifying a reference to the CS together with the value of the QoS attribute they can provide the service with at that time. If the negotiation is successful, then the AW can be instantiated with the CSs having the suitable QoS value, and the *Instantiated Workflow* (IW) represents the requested application ready to be executed.

In the proposed negotiation mechanism only SPs formulate new offers, and only the SC evaluates them. The rationale of this choice is twofold: on one hand it makes it possible to simulate what happens in a real market of services where an SC does not have enough information on the SPs strategies to formulate counteroffers; on the other hand it takes into account that the offers for a single functionality cannot be evaluated independently from the ones received for the other functionalities, i.e., negotiating over the attributes of the single AS cannot be done independently from each other. So, the negotiation mechanism should allow to negotiate with the SPs, and at the same time to evaluate the aggregated QoS value of the received offers for all the required functionality in the AW during the negotiation.

In order to meet these requirements, an iterative negotiation protocol [3], based on a Contract Net Iterated Protocol, is adopted. The negotiation occurs between the SC and the SPs available for each AS of the AW, and it may be iterated for a variable number of times until a *deadline* is reached or the negotiation is successful. Each iteration is referred to as a negotiation *round*, and the deadline is the number of allowed rounds. According to the protocol, at each negotiation round the SC sends $m*n$ call for proposals, where m is the number of ASs in the AW, and n is the number of SPs available to take part in the negotiation for each AS, and after waiting for the time set to receive offers (know as the *expiration time*), it checks first that there are offers for each AS; if not the SC declares a failure since it is not possible to instantiate the AW. Otherwise, it evaluates the received offers, and, according to the result of the evaluation (see Sec. 2.1), it starts another negotiation round, or it selects the best offers in terms of its own utility.

2.1 The Negotiation Evaluation

The SC evaluates the offers received at each negotiation round to check whether the global constraints specified by the user are met by using a solver of a Linear Programming problem so formulated. There are nm decision variables $x_{i,j}$ where i identifies one of the m ASs and j identifies one of the n SPs compatible with the i_{th} AS. The variable $x_{i,j}$ is equal to 1 if the j_{th} SP is selected for the i_{th} AS, 0 otherwise. Since only one SP has to be selected for each AS, then $\sum_{j=1}^{n} x_{i,j} = 1$ for all ASs.

In the general case of a multidimensional QoS $(Q_1, \ldots, Q_r)$, the n-tuple $(q_{i,j}^1, \ldots, q_{i,j}^r)$ of offered values is associated to each corresponding SP identified by the decision variable $x_{i,j}$. To check whether each QoS constraint is satisfied, taking into account all the ASs in the workflow, the aggregated values of the parameters $q_{i,j}^k$ offered by each selected SP must not exceed the user upper bound Q_k, i.e.:

$$aggrFunc_{i,k}\left(\sum_{j=1}^{n} x_{i,j} q_{i,j}^k\right) \leq Q_k, \forall k = 1, \ldots, r \tag{1}$$

where $aggrFunc$ depends on the type of the considered parameter. Typically, in the literature additive (e.g., price and execution time) and multiplicative (e.g., reliability and availability) parameters are studied [8], so $aggrFunc$ is either a summation or a multiplication over the number of ASs. Once solutions that satisfy the constraints are found, the SC evaluates their utility with the formula [1]:

$$U_{i,j}(SC) = \sum_{k=1}^{r} \frac{Q_{max(i,k)} - q_{i,j}^k}{Q_{max'(k)} - Q_{min'(k)}} \tag{2}$$

where $Q_{max(i,k)} = max(q^k_{i,j})$, $Q_{max'(k)} = aggrFunc_k(Q_{max(i,k)})$ aggregating the local maxima of the offers received for each AS, and $Q_{min'(k)} = aggrFunc_k(Q_{min(i,k)})$, aggregating the corresponding local minima.

Eq. 2 evaluates the SC's utility of an offered QoS value w.r.t. both the ones offered by the other SPs of the same service (local evaluation), and the QoS value of a possible instantiated workflow (global evaluation). In fact, $Q_{max(i,k)} - q^k_{i,j}$ gives an indication of how good the value of each QoS parameter is with respect to the QoS values offered by other SPs of the same AS (local evaluation) by taking as a reference the maximum offered value for that parameter. Local evaluation compared to $Q_{max'(k)} - Q_{min'(k)}$ gives an indication of how good the value of each parameter is with respect to the possible aggregated values of the same parameter for all the ASs (global evaluation).

The SC objective function is a maximization of the sum of the utilities for each $m-tuple$ of selected SPs that satisfies the QoS global constraints given by Eq.3:

$$max(\sum_{i=1}^{m}\sum_{j=1}^{n} x_{i,j}U_{i,j}(SC)) \tag{3}$$

2.2 The Negotiation Strategy

SPs strategies are modeled as a set of functions that are both time and resource dependent [4] taking into account both the *computational load* of the provider, and the *cost* of the provided service. The computational load accounts for the provider workload, i.e., the amount of service implementations it will deliver, while the cost of the service is directly proportional to its complexity.

For each SP the negotiation strategy is modeled by a Gaussian distribution that represents the probability distribution of the offers in terms of the provider's utility. As shown in Fig. 1, the mean value of the Gaussian *maxU* represents the best offer the SP may propose in terms of its own utility with the highest probability to be selected; while the standard deviation σ represents the attitude of the SP to concede during negotiation and it is given by $\sigma_{i,j} = maxU_{i,j} - maxU_{i,j}percent_{i,j}$, where $percent \in [0,1]$ represents the concession percentage of the SP with respect to its own utility. The parameter σ varies from SP to SP providing the same AS, so that the lower its computational load (in terms of available resources) is, the more it is available to concede in utility and the lower its reservation value is. The negotiation set for the SP is $[maxU - \sigma; maxU]$, where $maxU - \sigma$ is the reservation value.

In Fig. 1 the functions associated to two different SPs for the same AS are reported. The best offer is the same for both SPs (i.e. $maxU_1 = maxU_2$) since it is assumed that services providing the same functionality have the same utility value, while their concession strategies are different according to their workload when the negotiation takes place. In fact, σ_1 is greater than σ_2 meaning that SP_1 has a lower computational load than SP_2, so it concedes more in utility than SP_2.

At each negotiation round, the SP generates a new utility value corresponding to a new offer according to its Gaussian distribution (for values generated in the set

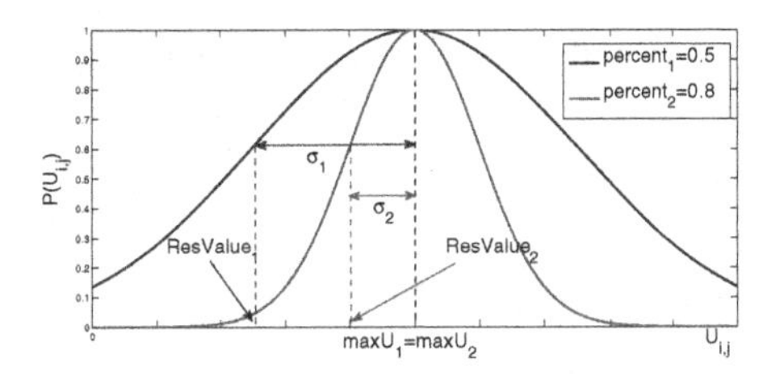

Fig. 1 An example of probability functions to compute new offers

$[maxU; maxU + \sigma]$, the values within the negotiation set $[maxU - \sigma; maxU]$ with the same probability to be selected, are considered): if this value is lower than the one offered in the previous round and within the negotiation set, then the SP proposes the new value; if this value is higher than the one offered in the previous round, or it is outside the negotiation set, the SP proposes the same value offered in the previous round. This strategy allows to simulate SPs that prefer not having a consistent loss in utility, even though by increasing the number of negotiation rounds the probability for the SP to move towards its reservation value increases.

3 A Cost-Based Testbed

A preliminary set of experiments was carried out in order to determine the main factors affecting the negotiation success/failure, and to evaluate the impact of the SPs negotiation strategies on the negotiation progress. The experiments were aimed at verifying the possibility to extract information that can be used by the SC to decide whether to iterate or to stop the negotiation according to the current situation.

In these experiments, the QoS attribute is the service price, so the QoS value of the requested application is additive in the number of ASs, and it does not depend on the structure of the AW. The utility value for the SP represents the price for the service it offers, so *maxU* is the highest price (*bestPrice*) offered by the SP, and it is the same for all SPs of the same AS. An SP offer is $Price_{i,j}$, and the reservation price for the SP is $bestPrice - \sigma$. It is assumed that the more complex a service functionality is, the higher its "market price" is, i.e. the variability in prices for different ASs is proportional only to their complexity. To simulate this variability, a parameter k is used: the more complex the functionality provided by a service is the higher the k value is. In particular, for SPs providing less complex services $k \leq 1$, while for SPs providing more complex services $k > 1$. The k parameter is equal for all SPs of the same AS, meaning that services providing the same functionality have the same market price. In fact, k determines the mean value of the Gaussian distribution, and so the *bestPrice* for an ASi is:

$$bestPrice_i = \frac{globalPrice \;\, k_i}{m} \qquad i \in [1, \ldots, m] \tag{4}$$

where, m is the number of ASs in the AW. So, Eq. 4 takes into account both the computational cost of the offered service, and also the assumption that the requested price *globalPrice* is not "unreasonable" compared to the market price of the required ASs in the AW. A *feasible* solution exists if the QoS constraint is satisfied, i.e. $\sum_{i=1}^{m} \sum_{j=1}^{n} x_{i,j} Price_{i,j} \leq globalPrice$.

If for each AS $k > 1$, then the QoS constraint is never satisfied at the first round.

3.1 Experimental Results

The impact of the σ parameter is evaluated considering a configuration with k and the number of AS (i.e. provider agents) fixed. The default price is assigned to each ASi according to Eq.4, with $m = 5$ and $k_i = 2.4, 2.0, 1.3, 1.0$ and 0.8 (average value of $k = 1.5$), approximately corresponding to 32%, 27%, 17%, 13% and 11% of the *globalPrice*. The corresponding SPs send as initial offer a price in the neighborhood of $bestPrice_i$ $[bestPrice_i - 5\%, bestPrice_i]$. The *percent* value randomly varies for each SP in the range [0.5, 1.0], so including the possibility to have SPs with the maximum computational load, not willing to concede (i.e., $percent = 1$). The maximum number of negotiation rounds is 100. In the case the SPs of AS3, AS4 and AS5 are not willing to negotiate, the negotiation always fails even with the minimum *percent* value for AS1 and AS2. Tables 1a, 1b, 1c, 1d report the average of the minimum *percent* values for each AS with respectively 2, 4, 8, 16 SPs.

Table 1 Minimum *percent* with respectively a) 2 SPs, b) 4 SPs, c) 8 SPs, d) 16 SPs for each AS

a)

Succ/Fail	%	AS1	AS2	AS3	AS4	AS5
successes	61	0.61	0.63	0.64	0.65	0.65
failures	39	0.74	0.74	0.74	0.70	0.69
successes	8	0.56	0.56	0.57	1	1
failures	92	0.67	0.68	0.68	1	1

b)

Succ/Fail	%	AS1	AS2	AS3	AS4	AS5
successes	95	0.59	0.60	0.61	0.61	0.60
failures	5	0.72	0.65	0.70	0.72	0.61
successes	35	0.54	0.56	0.57	1	1
failures	65	0.65	0.62	0.61	1	1

c)

Succ/Fail	%	AS1	AS2	AS3	AS4	AS5
successes	100	0.55	0.55	0.54	0.56	0.55
failures	0	-	-	-	-	-
successes	77	0.54	0.55	0.55	1	1
failures	23	0.60	0.61	0.59	1	1

d)

Succ/Fail	%	AS1	AS2	AS3	AS4	AS5
successes	100	0.53	0.53	0.52	0.52	0.52
failures	0	-	-	-	-	-
successes	100	0.52	0.53	0.53	1	1
failures	0	-	-	-	-	-

As expected, with SPs for AS4 and AS5 not willing to negotiate, the minimum *percent* values for the providers of the remaining ASs have to be lower than the ones obtained for configurations where all the SPs are available to make concessions. In fact, for successful negotiations, the average minimum value of *percent* is 0.55, while with an average minimum value of $percent = 0.63$ negotiation failures are obtained (third and fourth rows in tables 1a, 1b and 1c). The only exception is table 1d) where, increasing the number of SPs to 16 for each AS, a 100% rate of successes is obtained with an average minimum value of *percent* equal to 0.53. Moreover, in the case of the providers for AS4 and AS5 not willing to negotiate, there are more failures than successes in tables 1a and 1b, while more successes

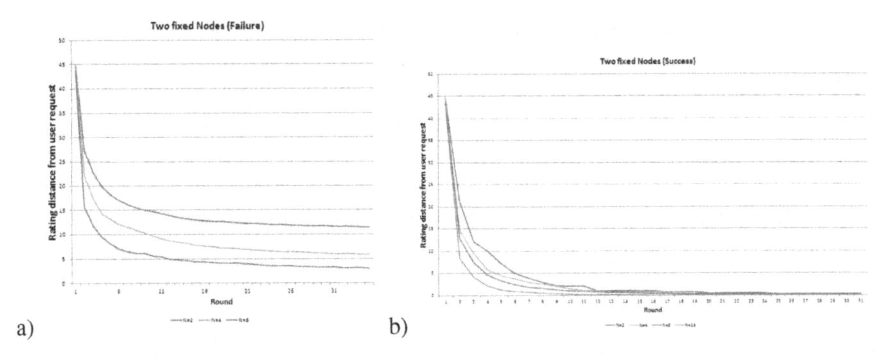

Fig. 2 Distance in case of failures (a) and successes (b)

than failures in table 1c. So, by increasing the number of SPs, the chances to succeed in the negotiation increase. In the case all SPs are willing to negotiate (first two rows in the tables) more successes than failures are obtained with smaller medium minimum values of *percent*. Finally, the higher the computational cost of the AS is, the smaller the corresponding average minimum value of *percent* is to have a success. This shows that the SC should negotiate essentially with SPs providing higher computational cost services since they impact more the negotiation success/failure.

In Fig. 2 the distance of the price obtained at each negotiation round from *globalPrice* is reported varying the number of available SPs given by:

$$\frac{((\sum_{i=1}^{m}\sum_{j=1}^{n} x_{i,j} Price_{i,j}) - globalPrice)100}{globalPrice} \tag{5}$$

In Fig. 2a the distance for the failure cases is plotted showing that the curve trend is the same varying the number of SPs. After the 25th round, the distance from the required price varies very little (0.1%) and there is a failure at the 100th round. This information can be used at runtime to dynamically set the negotiation deadline (e.g. in our experiments it can be 30 rounds). In Fig. 2b the distance for the success cases is plotted, and it shows that negotiation always ends before the 30th round.

4 Conclusions

Several efforts have been carried out in the areas of QoS-based service selection for SBAs. Some works propose algorithms to select service implementations relying on the optimization of a weighted sum of global QoS parameters as in [8] by using integer linear programming methods. In [2] local constraints are included in the linear programming model used to satisfy global QoS constraints. In [1] Mixed Integer Programming is adopted to find the optimal decomposition of global QoS constraints into local constraints so the best services satisfying the local constraints can be found. Typically, these works rely on static approaches assuming that QoS parameters of each service do not change during the selection process. Other approaches rely on negotiation mechanisms to select services according the QoS values [5, 6], but usually negotiation is carried out for each required service independently from the others. Attempts to propose a coordinated negotiation with all the providers of the different service in a composition have been proposed [3], but not empirical evaluation is provided to show that if a solution exists it is found.

This work proposes an approach for QoS-based service selection that takes into account the variability of service providers provision strategy. The use of an iterative negotiation mechanism allows to address the limitations of assuming static QoS values that is not realistic in market-based service scenarios, since providers might change dynamically their provision strategies according to market trends during service selection to become more competitive in the market during negotiation. For this reason, the service compositor negotiates with all available providers so to not discharge providers that may become more competitive during negotiation. The proposed mechanism allows to evaluate the progress of the negotiation, so it can be stopped if the SC utility is not improving. This feature is useful in service-based application settings since negotiation is computationally expensive. Furthermore, aggregated offers are evaluated at each negotiation round since the selection of one service cannot be done independently from the other services. This is even more crucial in case of multidimensional QoSs.

References

1. Alrifai, M., Risse, T.: Combining global optimization with local selection for efficient qos-aware service composition. In: Proceedings of the 18th International Conference on World Wide Web, WWW 2009, pp. 881–890. ACM, New York (2009), `http://doi.acm.org/10.1145/1526709.1526828`, doi:10.1145/1526709.1526828
2. Ardagna, D., Pernici, B.: Adaptive service composition in flexible processes. IEEE Trans. on Software Eng. 33(6), 369–384 (2007)
3. Di Napoli, C.: Using software agent negotiation for service selection. In: Mele, F., Ramella, G., Santillo, S., Ventriglia, F. (eds.) BVAI 2007. LNCS, vol. 4729, pp. 480–489. Springer, Heidelberg (2007)
4. Faratin, P., Sierra, C., Jennings, N.R.: Negotiation Decision Functions for Autonomous Agents. International Journal of Robotics and Autonomous Systems 24, 3–4 (1998)
5. Paurobally, S., Tamma, S., Wooldridge, M.: A framework for web service negotiation. ACM Trans. on Autonomous and Adaptive Systems 2(4) (2007)
6. Siala, F., Ghedira, K.: A multi-agent selection of web service providers driven by composite qos. In: IEEE Symposium on Computers and Communications, pp. 55 –60 (2011)
7. Strunk, A.: Qos-aware service composition: A survey. In: Proceedings of IEEE 8th European Conference on Web Services, pp. 67–74. IEEE Computer Society (2010)
8. Zeng, L., Benatallah, B., Ngu, A.H., Dumas, M., Kalagnanam, J., Chang, H.: Qos-aware middleware for web services composition. IEEE Trans. on Soft. Eng. 30(5), 311–327 (2004)

D3S – A Distributed Storage Service

Rui Pedro Lopes and Pedro Sernadela

Abstract. The Internet growth allowed an explosion of service provision in the cloud. The cloud paradigm dictates the users' information migration from the desktop into the network allowing access everywhere, anytime. This paradigm provided a adequate environment to the emergence of online storage services, such as Amazon S3. This kind of service allows storing digital data in a transparent way, in a pay-as-you-go model. This paper describes an implementation of an S3 compatible cloud storage service based on peer-to-peer networks, in particular, through the BitTorrent protocol. This approach allows taking advantage of the intrinsic features of this kind of networks, in particular the possibility for simultaneous downloading of pieces from different locations and the fault tolerance.

Keywords: Storage, BitTorrent, Amazon S3, Cloud Computing, P2P.

1 Introduction

Cloud computing has emerged as an important paradigm for managing and delivering services over the Internet. The term "Cloud" comes from the data center hardware and software association and is usually based on a model of payment associated with the use of resources. This on-demand, pay-as-you-go model creates a flexible and cost-effective means for using and accessing distributed computing resources [12, 1, 10].

Several providers have appeared, such as Amazon, Hewlett-Packard, IBM, Google, Microsoft, Rackspace, Salesforce, on the promise to increase revenue by optimizing their existing IT infrastructure and personal. According to the abstraction layer as well as to the type of service, users' are allowed to access applications on the cloud (SaaS), to deploy applications to the cloud without the added

Rui Pedro Lopes · Pedro Sernadela
Polytechnic Institute of Bragana, Bragana, Portugal
e-mail: {rlopes,psernadela}@ipb.pt

J.B. Pérez et al. (Eds.): *Trends in Prac. Appl. of Agents & Multiagent Syst.*, AISC 221, pp. 17–24.
DOI: 10.1007/978-3-319-00563-8_3

complexity of software and hardware management (PaaS), and access to "low level" storage and computing capabilities (IaaS) [9].

This paradigm suffers, essentially, from two drawbacks. First, the centralized nature is vulnerable to single point of failure, originated from several causes, such as fire, lightning storms, hardware failure, and others [3, 17, 11]. Second, it is necessary a considerable investment on the infrastructure and a lot of experience in Information Technologies (IT). Industry leaders like Amazon and Google take advantage on the fact that they already possess large infrastructures and knowledge to fuel their business (Amazon S3 and Google Drive respectively). However, guarantying 99.99% availability requires a huge investment that usually companies resist to make.

This paper describes a first approach to address centralization related problems in cloud services, specifically on storage IaaS clouds. We will use Amazon S3 as base, instantiating an S3 service over a BitTorrent network. In this model, we store objects in BitTorrent swarms, as a means to achieve throughput, redundancy, scalability, capacity, and others. Many of these requirements are naturally available in peer-to-peer networks, although dependent on the number of peers that belong to a specific swarm, as well as on node availability. This implementation will be used in the future to assess availability and throughput.

2 Storage on the Cloud

Cloud computing is the result of the evolution of different technologies that allowed more processing capacity, virtualization of resources and the ability for data storage in the network. The combination of these factors led to a new business paradigm that responds to a set of problems, such as scalability, cost reduction, fast operation of applications and solutions [15].

According to the degree of service abstraction, it is possible to distinguish two different architectural models for clouds [4]. One is designed to allow the expansion of computing instances as a result of the increase on the demand (Software-as-a-Service and Platform-as-a-Service). The other model is designed to provide data and compute-intensive applications via scaling capacity – Infrastructure-as-a-Service. Regarding the deployment model, a cloud may be restricted to a single organization or group (private clouds), available to the general public over the Internet (public clouds), or a composition of two or more clouds (hybrid clouds) [4, 10, 14, 18].

Among the multitude of services available on the cloud, storage services are rather popular. Services such as Dropbox[1], Amazon S3[2], Google Drive[3] and others, provide an easy to use, flexible way to store information and easy access anytime, anywhere.

---

[1] http://www.dropbox.com

[2] http://aws.amazon.com/s3

[3] https://drive.google.com

2.1 Amazon S3

The Amazon Simple Storage Service (Amazon S3) is an online service that enables storing data in the cloud. This service is designed to provide 99.999999999% durability and 99.99% availability of objects over an year. Amazon S3 adopts the pay-per-use model: the prices are based on the data location (region) and the request count (GET, PUT, COPY, POST, and LIST) and the data transfers into and out of an Amazon S3 region.

Data is organized in two levels. At the top level, there is the concept of buckets, similar to folders, identified by unique global name. Buckets are needed to organize the Amazon S3 namespace and to identify the account responsible for storage and data transfer charges.

Each bucket can store a large number of objects, each comprising data (blob) and metadata. The data portion is opaque to Amazon S3 and has a limit of 5 terabytes. The metadata is a set of name-value pairs that describe the object (Content-Type, Date, ...), with a limit of 2 kilobytes. An object is uniquely identified within a bucket by a *key* (name) and a *version ID*. The key is the unique identifier for an object within a bucket. Users can create, modify and read objects in buckets, subject to access control restrictions.

When users register an Amazon Web Services (AWS) account, AWS assigns one Access Key ID (a 20-character, alphanumeric string) and one Secret Access Key (a 40-character string). The Access Key ID uniquely identifies an AWS Account. AWS uses a typical implementation that provides both confidentiality and integrity (through server authentication and encryption).

These kind of service, typically provide a specific, SOAP or REST based, interface. The S3 REST API[4] uses a custom HTTP scheme based on a keyed-HMAC (Hash Message Authentication Code) for authentication. Standard HTTPAuthorization header is used to pass authentication information.

To authenticate a request, selected elements of the request are concatenated to form a string with the following form: "AWS AWSAccessKeyId:Signature". In the request authentication, the first "AWSAccessKeyId" element identifies the user that make the request and the secret key that was used to compute the "Signature". The "Signature" element is the SHA1 hash of the request selected elements [7]. In the Amazon S3 Request authentication this algorithm takes as input two byte-strings: a key and a message. The key corresponds to the AWS Secret Access Key and the message is the UTF-8 encoding of one string that represents the request. This string includes parameters like the HTTP verb, content MD5, content type, date, etc.. that will vary from request to request. The output of HMAC-SHA1 is also a byte string, called the digest. The final "Signature" request parameter is constructed by Base64 encoding this digest. When the system receives an authenticated request, it fetches the AWS Secret Access Key and uses it in the same way to compute a "Signature" for the message it received. It then matches signatures to authenticate the message.

[4] `http://docs.amazonwebservices.com/AmazonS3/latest/dev/RESTAuthentication.html`

The S3 architecture is designed to be programming language-neutral providing REST and a SOAP interfaces to store and retrieve objects. REST web services were developed largely as an alternative to some of the perceived drawbacks of SOAP-based web services [6]. With REST, standard HTTP requests are used to create, fetch, and delete buckets and objects. So, this interface works with standard HTTP headers and status codes (Table 1).

Table 1 HTTP operations with URI pattern that can be performed in S3

Operations	Description
GET /	get a list of all buckets
PUT /{bucket}	create a new bucket
GET /{bucket}	list contents of bucket
DELETE /{bucket}	delete the bucket
PUT /{bucket}/{object}	create/update object
GET /{bucket}/{object}	get contents of object
DELETE /{bucket}/{object}	delete the object

3 Distributed S3 Storage Service

The low-level usage patterns of Amazon S3 are, essentially, storing, updating and retrieving sequence of bytes through SOAP or REST WS, having in mind the durability and availability defined in the SLA. In this pattern, the access speed depends on the end-to-end throughput to where buckets and objects are stored (Amazon's data centers).

We implemented the S3 REST services as a loop-back to a BitTorrent swarm, providing locally the same functionality as Amazon's interface – the Distributed S3 Storage Service (D3S). From now on, we will identify the locally available gateway between S3 server-side interfaces and the BitTorrent swarm as a *D3S node*.

According to the usage patterns, the D3S node will receive requests through the REST WS and translate them to BitTorrent operations. Storing data will create a .torrent file and distribute it among a set of peers, that will define the swarm associated to these pieces of data. Updating data will require that the corresponding piece on the file to be updated. Retrieving data will require that the peer will download and make available the corresponding file by contacting a set of peers from the same swarm.

In a typical, centralized, client-server architecture, clients share a single server. In this case, as more clients join the system, fewer resources are available to each client, and if the server fails, the entire network fail as well.

Peer-to-peer networks are intrinsically distributed, where each peer behaves as a client and as a server simultaneously. In other words, "*A distributed network architecture may be called a Peer-to-Peer (P-to-P, P2P,...) network, if the participants share a part of their own hardware resources (processing power, storage capacity, network link capacity, printers,...). These shared resources are necessary to provide the Service and content offered by the network (e.g. file sharing or share workspaces for collaboration): They are accessible by other peers directly, without passing intermediary entities*" [16].

3.1 Overall System Architecture

BitTorrent became the third generation protocol of P2P networks, following Napster and Gnutella, where the main difference between the previous generations is the creation of a new network for every set of files instead of trying to create one big network of files using servers. Nowadays, BitTorrent is actually one of the most popular protocols for transferring large files, with over 150 million active users[5].

Storing and organizing objects in D3S requires the creation and distribution of files in a set of BitTorrent peers. Each object, in the context of S3, will correspond, in our architecture, to a single file, which is registered in the swarm through a shared `.torrent`.

The overall architecture is depicted in Figure 1. The file that corresponds to an object is created in the local host, through the interface provided by a D3S node. The file is sliced up in pieces by the BitTorrent peer, to get small amounts of verifiable data from several peers at a time. Each time a piece is fully downloaded, it will be checked (using the SHA1 algorithm). The pieces are then spread in the swarm, following the "tit-for-tat" characteristic of BitTorrent [2].

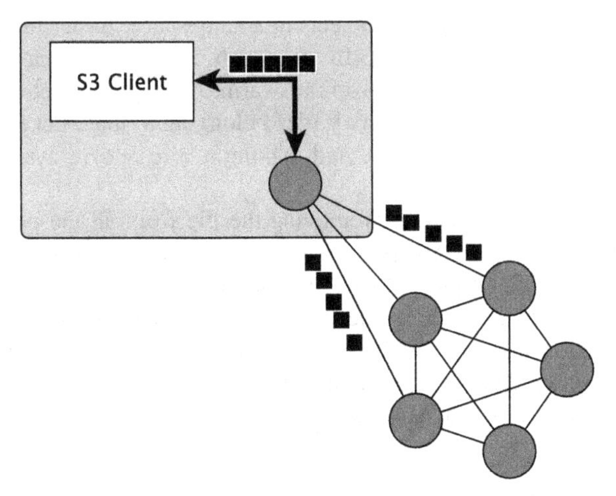

Fig. 1 Overall architecture implemented

[5] `http://www.bittorrent.com/intl/es/company/about/ces_2012_150m_users`

Clients involved in a torrent cooperate to replicate the file among each other using swarming techniques. A user who wants to upload a file first creates a torrent descriptor file (with the extension `.torrent`) that has to be made available to the other peers – this operation is called seeding.

Peers that want to download the file must first obtain the associated torrent file and connect to the specified tracker, which makes it possible to connect to other peers to download the pieces of the file (typically 256 KB but other piece sizes are possible too).

The tracker is not involved in the distribution of the file, it only coordinates the file distribution by keeping track of the peers currently active in the torrent – the tracker is the only centralized component of BitTorrent.

Pieces are typically downloaded non-sequentially and are rearranged into the correct order by the BitTorrent client. Those peers download it by connecting to the seed and/or other peers that has the file – acting as leechers.

Active peers report their state to the tracker and each time a peer obtains a new piece, it informs all the peers it is connected to. Interactions between peers are based on two principles [8]. First, the *choke algorithm* that encourages cooperation among peers and limits the number of peers a client is simultaneously sending data. Second, the *rarest first algorithm* that controls the pieces a client will actually request in the set of pieces currently available for download. The more peers join the swarm, the access speed and availability increases, contributing to the overall user experience [13].

3.2 Bucket and Object management

The role of a bucket is to organize objects in a namespace. In our implementation, this translates to a file name prefix. In turn, each object is associated to a single file, that will be uploaded to the BitTorrent swarm. Adding a bucket corresponds to defining a prefix that will be used in all files belonging to that bucket. Adding and object corresponds to creating a file and making a `.torrent` available to other peers.

Removing a file corresponds to removing the file from all the peers and eliminating the `.torrent`. This is performed through the User Manager and, when signaled, the peer will eliminate them consequently.

D3S nodes contact the tracker when needed, to manage the overlay network. In our implementation, each peer will also contact the User Manager, which maintains a database of user credentials, to authenticate each user's peers. The User Manager also maintain a relation of each user's `.torrent` files, residing in the `.torrent` repository.

The `.torrent` Manager receives `.torrent` files from the peers and associates them with the user that originated the request. `.torrent` files are generated when the user uploads a file to the S3 service. Peers will periodically pool the `.torrent` Manager to check for new additions, so that each one of them can replicate the file, thus making it available to the other peers as well as locally. Anytime that the user requests a file, it is already available locally. If the file is not available locally, it is downloaded from several replicas simultaneously, increasing the access speed.

As stated, we took advantage of RMI that provides the mechanism by which the server and the client communicate and pass information. Then we show the remote interface methods that can be invoked by the User Manager and the Torrent Manager to interact with the tracker.

This approach constitutes a bottleneck and a single point of failure. We intend, in the future, to move the User Manager to a Distributed Hash Table (DHT), to provide scalability and fault tolerance.

In sum, BT networks natively supports resource sharing, which requires self organization, load balancing, redundant storage, efficient search of data items, data guarantees, trust and authentication, massive scalability properties and fault-tolerance (i.e., if one peer on the network fails the whole network is not compromised) [5]. These characteristics are thus imported to our implementation, meeting many of the requirements of cloud storage services. Moreover, it intrinsically copes with scalability, redundancy and availability: the more replicas there are, the higher the redundancy, access speed and availability.

4 Conclusions and Future work

With the emergence of cloud computing, many cloud storage services have been offered and provisioned to customers. The pay-per-use model is, nowadays, an attractive solution to the traditional storage solutions. One well known example is Amazon S3, providing a transparent, fault tolerant and scalable storage service.

On the other hand, the BitTorrent protocol has made a big agitation on the file sharing community, and is nowadays one of the most used protocol in Internet. The advantages of this protocol are well know, particularly in terms of supporting distributed storage and efficient transfer of large files. The possibility to download (and upload) a file from several locations simultaneously allows excellent transfer speeds and the replicas allow for good availability.

Integrating an S3 compatible REST interface to a BitTorrent network allows using the scalability, openness and transfer speed in S3 compatible applications, without the need to rely on a single network connection to the provider.

In future work, we will investigate the relative performance of this approach to the cloud based service.

References

1. Akioka, S., Muraoka, Y.: HPC Benchmarks on Amazon EC2, pp. 1029–1034. IEEE, doi:10.1109/WAINA.2010.166
2. Cohen, B.: The bittorrent protocol specification, `http://www.bittorrent.org/beps/bep_0003.html` (accessed on January 14, 2013)

3. Cohen, R.: Cloud computing forecast: Cloudy with a chance of fail, http://www.forbes.com/sites/reuvencohen/2012/07/02/cloud-computing-forecast-cloudy-with-a-chance-of-fail/ (accessed on January 2, 2013)
4. Dikaiakos, M.D., Katsaros, D., Mehra, P., Pallis, G., Vakali, A.: Cloud Computing: Distributed Internet Computing for IT and Scientific Research. IEEE Internet Computing (5), 10–13 (2009), doi:10.1109/MIC.2009.103
5. Lua, E.K., Crowcroft, J., Pias, M., Sharma, R., Lim, S.: A survey and comparison of peer-to-peer overlay network schemes. IEEE Communications Surveys & Tutorials (2), 72–93 (2005), doi:10.1109/COMST.2005.1610546
6. Hamad, H.: Performance evaluation of restful web services for mobile devices. International Arab Journal of e-Technology
7. Krawczyk, H., Bellare, M., Canetti, R.: HMAC: Keyed-Hashing for Message Authentication. RFC 2104 (Informational) (1997), http://www.ietf.org/rfc/rfc2104.txt, Updated by RFC 6151
8. Legout, A., Urvoy-Keller, G., Michiardi, P.: Understanding bittorrent: An experimental perspective. INRIA Sophia Antipolis/INRIA
9. Marston, S., Li, Z., Bandyopadhyay, S., Zhang, J., Ghalsasi, A.: Cloud computing - The business perspective. Decision Support Systems (1), 176–189, doi:10.1016/j.dss.2010.12.006
10. Mell, P., Grance, T.: The NIST definition of cloud computing (draft), p. 145. NIST special publication
11. Ngak, C.: Gmail, google drive, chrome experience outages, http://www.cbsnews.com/8301-205_162-57558242/gmail-google-drive-chrome-experience-outages/ (accessed on January 2, 2013)
12. Ostermann, S., Iosup, A., Yigitbasi, N., Prodan, R., Fahringer, T., Epema, D.: A performance analysis of EC2 cloud computing services for scientific computing. Cloud Computing, 115–131
13. Parameswaran, M.: P2P networking: an information sharing alternative. Computer
14. Ramgovind, S., Mm, E., Smith, E.: The Management of Security in Cloud Computing. Security (2010)
15. Rimal, B.P., Choi, E., Lumb, I.: A Taxonomy and Survey of Cloud Computing Systems, pp. 44–51. IEEE, doi:10.1109/NCM.2009.218
16. Schollmeier, R.: A definition of peer-to-peer networking for the classification of peer-to-peer architectures and applications. In: Proceedings of thePeer-to-Peer Computing (2001)
17. Weinstein, N.: Netflix outage mars christmas eve, http://news.cnet.com/8301-1023_3-57560784-93/netflix-outage-mars-christmas-eve/ (accessed on January 2, 2013)
18. Zhang, Q., Cheng, L., Boutaba, R.: Cloud computing: state-of-the-art and research challenges. Journal of Internet Services and Applications 1(1), 7–18 (2010)

Evaluation of the Color-Based Image Segmentation Capabilities of a Compact Mobile Robot Agent Based on Google Android Smartphone

Dani Martínez, Javier Moreno, Davinia Font, Marcel Tresanchez,
Tomàs Pallejà, Mercè Teixidó, and Jordi Palacín

Abstract. This paper presents the evaluation of the image segmentation capabilities of a mobile robot agent based on a Google Android Smartphone. The mobile agent was designed to operate autonomously in an environment inspired in a soccer game. The pixel-based color image segmentation capabilities was performed by testing different two-dimensional look up tables created from two-dimensional color histograms of the objects appearing in the scenario. The best segmentation alternative was obtained when using of a two-dimensional look up table based on the H (hue) and V (value) color image description. The final conclusion is that a Google Android Smartphone has enough potential to define an autonomous mobile robot agent and play a game that requires color-based image segmentation capabilities.

Keywords: segmentation, mobile robot agent, Android, Smartphone.

1 Introduction

The development of a mobile robot ready to play a game is an application that requires the application of challenging technologies. The main goal of this work is the evaluation of the color-based image segmentation capabilities of a compact mobile robot based on Google Android Smartphone. The application scenario is

Dani Martínez · Javier Moreno · Davinia Font · Marcel Tresanchez ·
Tomàs Pallejà · Mercè Teixidó · Jordi Palacín
Computer Science and Industrial Engineering Department,
University of Lleida, 25001 Lleida, Spain
e-mail: {dmartinez,jmoreno,dfont,mtresanchez,tpalleja,
mteixido,palacin} @diei.udl.cat

J.B. Pérez et al. (Eds.): *Trends in Prac. Appl. of Agents & Multiagent Syst.*, AISC 221, pp. 25–32.
DOI: 10.1007/978-3-319-00563-8_4

inspired in the common game called soccer (in USA), football (in Europe), or futbol (in Spanish). This paper proposes the development of a comparative study on fast color-based image segmentation performed in a Google Android Smartphone. This paper has been developed in the direction of the creation of a full autonomous mobile robot agent based on this platform and with different potential game capabilities. In this paper, the mobile robot has to operate in a predefined scenario with the following color-based objects (Figure 1): floor, in green color; the field line limits, in white color; the ball, in a yellow main color; the goals, in red (red team) and blue (blue team) colors; and the mobile robots, also in red and blue colors. The main hypothesis is that current Smartphones have enough capabilities to perform fast color-based image segmentation and autonomously control the evolutions of a game oriented mobile robot.

Fig. 1 Image showing the ground field, the ball, one goal and two mobile robot agents

2 Background

The Android operating system is based on the Linux core and was developed by the Open Handset Alliance, led by Google. The creation of applications for Android based Smartphones and Tablets can be performed with the free Android SDK [1] that is based on the high level Java language with other specialized functions optimized for the specific hardware written in C. The main difference with the standard Java programming style is the use of events instead of a classic main function. The Integrated Development Environment (IDE) used in this work was the Eclipse 3.7.1 Indigo [2], the most popular IDE for Android which also includes the Java developer package.

2.1 Soccer Mobile Robots

The Robocup [3] domain and their leagues are the most popular reference for mobile robots which are designed to play a team game. This game domain define robots sizes and designs and up to 8 colors to define objects. Also it has different leagues; some of them are Small Size league, Medium Size league and Humanoid league. Nevertheless, the fulfillment of the rules defined by the Robocup domain is not the main goal of this work. Instead, we have defined a simplified game environment (Figure 1) containing one ball, two robots and two goals.

2.2 Fast Color-Based Image Segmentation

The image segmentation techniques applied in this paper are inspired in [4] which proposed an optimized threshold algorithm applied to the YUV color space to perform image segmentation and object color classification with a limited hardware; and in [5] and [6] which proposed the use of a YUV/RGB LUTs or other derived color spaces for color indexing based on specific calibration techniques to improve color segmentation and avoid the problems caused by lightning conditions. In this direction, this paper proposes the use of a two-dimensional LUT to define a fast segmentation methodology in order to take advantage of the limited number of colors defined in the proposed game.

3 Mobile Robot Implementation

Figure 1 shows an image of the mobile robot implementation based on a Google Android Smartphone. The mobile robot is a cylindrical box that supports the Smartphone and an additional battery and also contains the wheels, DC motors, and one electronic board to control the motors. The rear camera of the Smartphone is used to obtain images of the soccer playfield in low resolution (320x240 pixels) and high color quality. In this mobile robot the Smartphone analyzes the images acquired and sends basic commands: forward, back, turn left, turn right to the electronic board to drive the evolutions of the DC motors of the mobile robot. The additional battery is used to power directly the DC motors, the electronic board, and the Smartphone trough the USB connection. Then, the Smartphone can operate even when the external battery of the mobile robot is fully discharged, for example, to send a message to request a battery change.

The Smartphone used in this proposal is the HTC Sensation powered by a dual-core 1.2 GHz processor, 768 MB of RAM memory, and Android 4.0.3. The Smartphone has several embedded sensors: GPS, ambient light sensor, digital compass, three-axial accelerometer, gyroscope, multi-touch capacitive touch screen, proximity sensor, microphone, and a frontal and a rear camera. An important feature of one Smartphone is its high resolution screen that will be very useful to show some real-time information of mobile robot internal operation. The Smartphone device runs an agent APP which includes three basic capabilities: the main functional threads, the graphical interface, and the sensor managers. There are many specific threads defined, one thread has to receive the camera images and implement different color-based segmentation functions, two threads sends and receives high-level movement orders with the electronic board that control the motors of the mobile robot. The definition of different threads in Android requires the initialization of an instance of extended thread class and override the run() function which contains the code that the thread will execute. Normally this function contains the main code with the computational load of the application.

3.1 Image Processing Implementation

Once the camera of the Smartphone is initialized a preview callback function is called each time that a new image is generated. This function is uses to implement the color-based image segmentation capabilities of the mobile robot agent. Figure 2 shows the execution timeline required by the preview callback function to process the current image. The blue line in the figure shows an irregular behavior for the frame processing time. This effect is caused by the internal implementation of the Android camera functions and probably by other background applications running simultaneously in the Smartphone. After a careful analysis, the cause of the problem was identified as the Android's garbage collector module that operates very frequently to dynamically allocate different memory requested.

This problem can be solved on Android version 2.2 (or later) by using a new callback implementation *setPreviewCallbackWithBuffer* that avoids memory demands for each frame using the same static buffer to allocate image data, and thus, preventing frequent calling to the garbage collector (see Fig. 2 red line).

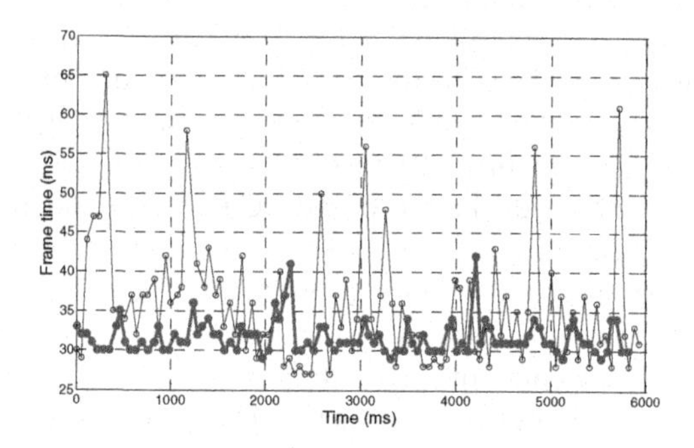

Fig. 2 Time lapse between consecutive calls to the preview callback function (blue line) and to the preview callback function with static buffer (red line)

3.2 Influence of the Image Color Space

The development of this work requires the use of different color spaces to evaluate with one has the best segmentation performances when applying a two dimensional LUT to classify the colors of the images. Figure 3 shows the histograms of the time lapse between consecutive calls to the preview callback function (with static buffer) for different color spaces. The red histogram shows the time lapse values when applying a LUT by using the original YUV color space image. This process is performed in most of the cases by spending 30 ms. The green histogram shows the time lapse spend when converting the original YUV (equivalent to YCbCr) color image into the RGB color space and then applying a LUT to classify the pixels of the image. This process is performed in most of the

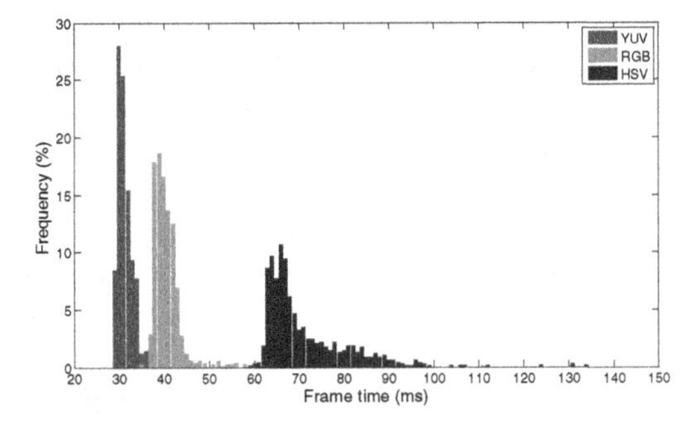

Fig. 3 Histograms of time lapse between consecutive calls to the preview callback function when classifying images in the YUV, RGB, and HSV color spaces

cases by spending 40 ms. Finally, the blue histograms showing the callback processing time when converting the original YUV image into the HSV color space and then apply a LUT to individually classify the pixels of the images. This process is performed in most of the cases by spending 67 ms. These poor results are caused by the difficult conversion between of the YUV and HSV image color spaces using high level procedures.

4 Fast Image Segmentation

The proposed methodology to perform fast image segmentation is the definition of a LUT for each specific color space used. In this analysis, the LUTs are computed off-line and the evaluation of the classification performances requires the development of the following steps: 1) acquire several images of the game playfield with the rear camera of the Smartphone, 2) manually label the different elements and objects of the images, 3) compute a two-dimensional histogram of all pixels of the same color, and 4) combine the histograms of the different objects in a two-dimensional LUT. Figure 4 also shows three examples of two-dimensional histograms corresponding to the ball and to blue and red color. These histograms have been computed by using the RG color spaces with 256 bins in each axis. These histograms can be joined in a unique LUT according the different pertinence probabilities of each object.

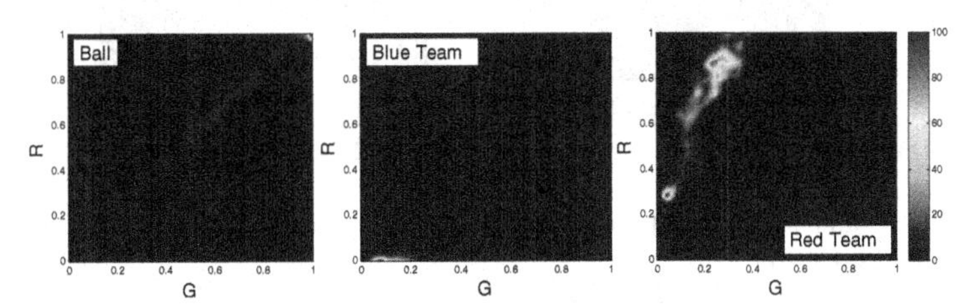

Fig. 4 Example RG histograms computed with the pixels corresponding to the ball, blue robot and goal team, and red robot and goal team (see figure 1)

Table 1 shows the results of a complete analysis performed to evaluate different color space combinations in the definition of a two-dimensional LUT optimized for the high quality images obtained with the Smartphone. The number of bins was limited to 32 after a trial and error test to simplify the final implementation in the Smartphone. The relative error shown in Table I is computed as the difference between a manually labeled image and the same image automatically classified with the LUT which was automatically obtained from the selected pixels of different images. The results in the column labeled as H have been obtained by creating the LUT from the conventional histograms and the results in the columns labeled as $\hat{H}$ have been obtained after convolving the histogram with a disk of 2 pixels radius (to expand the color-space area covered by each object in the histogram).

The analysis performed in this work has been developed by considering all combinations between the color spaces RGB, Gray, HSV, and YUV (labeled as Y-Cb-Cr). Additionally, the areas outside the game playfield field have been discarded and not compared. Figure 5-a shows the best LUT automatically obtained in this experiment, corresponding to the HV color spaces (case: histogram not convolved), and figure 5-b shows the image classification results obtained with this LUT. The average segmentation error obtained after comparing pixel by pixel the manual and automatic classification performed was 0.87%. The next best alternatives were based on the use of the HY, and the HG color spaces. Probably these results are strongly correlated with the high quality of the image acquired with the camera and all auto-adjusted photo-image features included in the Smartphone. The LUT shown in figure 5-a is very small (32x32 bins) and can be easily incorporated into the application of the Smartphone.

Figure 5-b also shows that there is an empty area in the LUT ($0.7<H<0.95$) that can be used, if required, to define a new color-based object (this area corresponds to the violet or purple color). Additionally, the results of figure 5-a also suggest that the basic classification information of the LUT could be expressed directly as different interval rules (as proposed in [4]) but it must be noticed that this alternative color classification will have always lower time-classification performances than the use of a LUT.

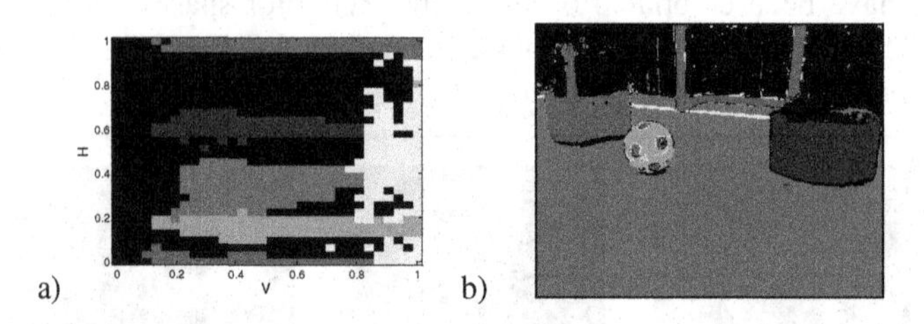

Fig. 5 Best segmentation LUT obtained in the HV color space (a) and its image classification results (b)

Table 1 Automatic image classification results

Color Spaces	Object segmentation relative error (%)										Average error (%)	
	Ball		*Field*		*Lines*		*Red Team*		*Blue Team*			
	H	$\hat{H}$	H	$\hat{H}$	H	$\hat{H}$	H	$\hat{H}$	H	$\hat{H}$	H	$\hat{H}$
R-G	1.18	1.38	3.34	3.43	0.91	0.92	1.15	0.73	1.63	1.68	1.64	1.63
R-B	5.35	7.55	2.33	2.20	0.69	0.65	5.74	7.56	0.94	0.85	3.01	3.76
R-H	0.41	0.77	2.19	2.08	0.79	0.73	0.56	0.54	0.57	0.51	0.90	0.93
R-S	2.84	4.41	2.67	2.45	0.71	0.66	3.35	4.25	1.26	1.13	2.17	2.58
R-V	2.16	2.48	2.97	3.07	0.90	0.93	1.88	2.42	1.26	1.26	1.83	2.03
R-Y	1.09	1.49	3.42	3.31	0.72	0.76	1.00	0.79	1.67	1.62	1.58	1.59
R-Cb	0.46	0.22	2.29	1.91	0.89	1.07	1.17	1.45	0.70	0.65	1.10	1.06
R-Cr	1.03	1.38	3.21	3.21	0.76	0.83	0.94	0.73	1.54	1.65	1.50	1.56
R-Gray	1.00	1.42	3.28	3.31	0.71	0.76	1.02	0.76	1.61	1.63	1.52	1.58
G-B	0.95	0.75	2.32	2.55	0.83	0.71	1.96	2.12	0.91	0.78	1.39	1.38
G-H	0.42	0.80	2.19	2.08	0.65	0.63	0.75	0.60	0.45	0.45	0.89	0.91
G-S	6.41	4.20	14.46	13.11	1.43	1.41	8.76	9.68	1.57	1.59	6.53	6.00
G-V	3.31	3.64	3.04	2.83	1.70	1.89	3.21	2.54	3.24	2.69	2.90	2.72
G-Y	3.26	2.75	3.61	3.73	1.38	1.39	0.91	0.73	2.31	2.80	2.29	2.28
G-Cb	1.08	0.27	5.50	6.26	0.63	0.60	4.94	5.13	0.72	0.54	2.57	2.56
G-Cr	1.00	1.46	3.03	3.19	0.78	0.73	1.01	0.73	1.33	1.43	1.43	1.51
G-Gray	4.40	3.04	3.97	3.80	1.95	1.97	1.00	0.72	2.21	2.80	2.71	2.47
B-H	0.41	0.77	2.25	2.04	0.66	0.56	0.75	0.61	0.41	0.45	0.90	0.89
B-S	17.20	20.73	17.28	19.03	0.77	0.93	5.34	5.54	0.81	0.68	8.28	9.38
B-V	5.61	7.37	2.07	2.92	0.83	0.72	5.78	7.34	0.58	1.62	2.97	3.99
B-Y	0.52	0.35	6.95	8.34	0.79	0.61	5.82	7.16	0.68	0.57	2.95	3.41
B-Cb	0.43	0.32	13.58	10.51	0.75	0.59	12.27	9.26	0.71	0.53	5.55	4.24
B-Cr	1.16	1.73	5.17	4.96	0.69	0.61	0.95	0.78	3.59	3.61	2.31	2.34
B-Gray	0.36	0.32	8.60	8.13	0.65	0.65	7.44	6.95	0.71	0.54	3.55	3.32
H-S	0.42	0.74	2.30	2.10	0.63	0.60	0.72	0.66	0.61	0.53	0.94	0.93
H-V	0.42	0.69	2.18	2.10	0.77	0.72	0.56	0.57	0.41	0.46	0.87	0.91
H-Y	0.42	0.78	2.20	2.08	0.68	0.67	0.71	0.61	0.45	0.47	0.89	0.92
H-Cb	0.39	0.29	14.85	14.95	13.78	14.94	0.48	0.65	0.48	0.44	6.00	6.25
H-Cr	0.36	0.51	2.86	2.51	1.74	1.79	0.57	0.53	0.44	0.50	1.19	1.17
H-Gray	0.43	0.73	2.24	2.10	0.67	0.68	0.72	0.61	0.48	0.47	0.91	0.92
S-V	5.09	7.44	4.24	5.17	1.43	1.12	4.75	5.43	1.30	1.34	3.36	4.10
S-Y	3.84	3.94	4.55	5.89	1.29	1.04	5.02	6.82	1.76	1.69	3.29	3.88
S-Cb	1.25	0.37	12.36	22.54	0.70	0.69	11.69	21.26	0.65	0.57	5.33	9.09
S-Cr	0.99	1.43	2.86	2.78	0.67	0.69	0.98	0.83	1.20	1.05	1.34	1.36
S-Gray	3.77	3.38	4.97	5.96	0.83	0.79	4.98	5.73	1.72	1.67	3.25	3.51
V-Y	2.58	2.51	3.30	3.03	0.91	1.17	2.57	2.17	1.76	1.56	2.22	2.09
V-Cb	0.55	0.51	5.68	5.92	0.85	0.86	4.75	5.11	0.71	0.58	2.51	2.60
V-Cr	1.11	1.36	7.85	13.43	0.83	0.80	0.94	0.73	6.18	11.60	3.38	5.58
V-Gray	3.64	3.16	3.91	3.03	1.53	1.76	2.57	2.13	1.64	1.52	2.66	2.32
Y-Cb	0.27	0.29	14.10	13.19	0.69	0.64	12.78	11.96	0.73	0.55	5.71	5.33
Y-Cr	1.07	1.42	3.31	3.68	0.73	0.74	0.91	0.75	1.53	2.01	1.51	1.72
Y-Gray	2.28	1.78	12.73	12.63	1.93	1.33	10.88	10.88	2.88	2.88	6.14	5.90
Cb-Cr	0.39	0.49	2.54	2.33	1.34	1.46	0.53	0.49	0.49	0.44	1.06	1.04
Cb-Gray	0.31	0.27	13.12	11.14	0.66	0.64	11.82	9.91	0.71	0.54	5.32	4.50
Cr-Gray	1.08	1.64	3.32	3.78	0.76	0.80	0.96	0.89	1.53	2.18	1.53	1.86

5 Conclusion

This paper presents the evaluation of the image segmentation capabilities of a compact mobile robot based on Google Android Smartphone. This proposal of a mobile robot takes advantage of the quality of the images provided by such devices. The time required to apply an image processing algorithm to the image acquired by the Smartphone has been analyzed and the effect of using a buffer in the preview callback function has been evaluated. Results show that the original YUV image can be classified in 30 ms whereas the converted RGB and HSV images in 40 and 67 ms respectively.

The best classification LUT was obtained from the two dimensional histograms of the H and V color spaces with an average error in the classification of the pixels of only 0.87%. The objection of using the HSV color space is that the conversion of the original images in the YUV color space to the HSV color space doubles the time required to classify an image. The time results obtained must depend largely on the Smartphone used but the time ratio is expected to be similar.

Future work will be focused in image calibration, algorithm optimization, and in additional image analysis procedures designed to estimate the relative distance to the different objects located in the playfield. Then the next efforts will be focused in the definition of the basic operational agent required to convert a Google Android Smartphone into a gaming mobile robot.

References

1. Android Developers, `http://developer.android.com`
2. The Eclipse Foundation open source community website, `http://www.eclipse.org`
3. Robocup, `http://www.robocup.org`
4. Bruce, J., Balch, T., Veloso, M.: Fast and inexpensive color image segmentation for interactive robots. In: IEEE/RSJ International Conference on Intelligent Robots and Systems (IROS 2000), vol. 3, pp. 2061–2066 (2000)
5. Neves, A.J.R., Pinho, A.J., Martins, D.A., Cunha, B.: An efficient omnidirectional vision system for soccer robots: From calibration to object detection. Mechatronics 21(2), 399–410 (2011)
6. Jamzad, M., Sadjad, S.B.S., Mirrokni, V.S., Kazemi, M., Chitsaz, H.R., Heydarnoori, A., Hajiaghai, M.T., Chiniforooshan, E.: A Fast Vision System for Middle Size Robots in RoboCup. In: Birk, A., Coradeschi, S., Tadokoro, S. (eds.) RoboCup 2001. LNCS (LNAI), vol. 2377, pp. 71–80. Springer, Heidelberg (2002)

Distributed and Specialized Agent Communities

Jesús Ángel Román, Sara Rodríguez, and Juan Manuel Corchado

Abstract. SCODA (Distributed and Specialized Agent COmmunities) is presented, like a new modular architecture for multi-agent systems development. By means of SCODA, multi-agent systems development is allowed under specialized modular philosophy, through it, the functionalities of the system can be extended in scaled form, according to the objectives. SCODA is composed by small subsystems of agents called, Specialized Intelligent Communities (CIE), which provide the necessary functionalities to solve the objectives needed across distributed services. By means of these CIE, scalability of the systems is allowed, so that they could be re-used in different developments, independently of his purpose.

Keywords: multi-agent systems, virtual organizations, dynamic architectures, specialization.

1 Introduction

The development of increasingly complex systems entails the need for develop capable components to be reused, so that the functionality that they provide can be used in other developments, keeping a compatibility between them. This philosophy is one that follows the object-oriented technology [1][2], where objects are encapsulated independently and can be reused in different developments with very different purposes, giving developers a significant advantage in terms of time spent in developing the system. This approach focused on Multi-agent Systems (Multi-Agent Systems, MAS) involves that, the development of an application based on multi-agent systems can be used in other developments, even with the overall purpose very different. For carrying out this approach, we need think that the set of agents that form a multi-agent system has to keep certain standardization and size, to be reused.

Jesús Ángel Román · Sara Rodríguez · Juan Manuel Corchado
University of Salamanca, Salamanca, Spain
e-mail: {zjarg,srg,corchado}@usal.es

J.B. Pérez et al. (Eds.): *Trends in Prac. Appl. of Agents & Multiagent Syst.*, AISC 221, pp. 33–40.
DOI: 10.1007/978-3-319-00563-8_5

The concept of organization has been studied extensively in sciences such as economics, sociology and psychology, and have also been several authors who have applied the concept of organization to development of multi-agent systems [3][4][5]. Applying the concept of organizing on multi-agent systems further increases the efficiency of itself having in account there is a check on the objectives to be met, both individual and collective, are established norms of behavior among agents, somehow there is a tasks division or specialization, that is, the function of the system is normed [6][7][8][9][5] If it also wants that a multi-agent system has the ability to be used in different developments regardless of their overall purpose, it would be necessary to define a new modular architecture and specialized, based on some sort of minimum size organization that establishes coordination norms between agents. The architecture would have a high degree of specialization at the organizational level, because the architecture of a multi-agent system determines the composition of the system itself, and the mechanisms used by agents to interact with their environment [10].

The main idea is design and implement a standard multi-agent architecture that meets the necessity of be reused in several developments in an efficient way. This architecture is based on a philosophy of business organization to improve the integration and development of multi-agent systems, where it is improved the distribution of the agents involved in the system and the services they implement. For this reason, is introduced the concept of community of agents as an organizational unit, and the concept of specialization applied thereto in order to obtain an optimization in the management and implementation of the system to be developed.

The article is structured as follows: Section 2 introduce several concepts, important for this work. Section 3 introduces SCODA, the architecture developed in this work. Finally, some experimental results and conclusions are given in Section 4 and 5.

2 Related Work

A multi-agent system is composed by a group of agents which act as some kind of organization. The main properties to characterize the internal behavior of the agents are: the type of reasoning and how they act, for example, reactive, model-based, based on goals. Its adaptability, its perception and characterization the environment in which they are situated, including its computer infrastructure and their relationships with other agents, and the degree of autonomy in the actions they take [29]. The organization is used to describe this group of agents that are coordinated through a series of patterns and the establishment of some roles to achieve the objectives of the system. There are various approaches that exist in this area [6][11][4][7][8], depending on the structure and internal organization of the organizational structure. The behavioral, the objectives and the interaction of the agents are features that mark its global behavior and the aims.

The concept of community differs from concept of organization that is more spontaneous and natural. The community relies on the common interests of those who work in a facility, and to carry out its regulation requires a system of co-management contract [12][13] [14] In the work of [15][16] the term community is used to define a group of heterogeneous agents. Another agent community approach is given in [11], which defines the structure of the Cougaar agents.

This paper presents the definition of Intelligent Communities (CI). These communities, adopt features defined in other organizations of agents besides the features defined in Cougaar. Intelligent Communities are composed of a small number of agents, so that the communication between the agents is multidirectional, and not entail added computational effort. Decision making is centralized in decisions relative of whole community and will be independent for each agent in the work they perform individually, always to the benefit of the community. For this there is a formalization of the tasks thus providing a tool for agents for individual decision making. In an Intelligent Community there will be a two-level hierarchy as to the work of community control. One agent will be involved in the control over the work of other members having account as based a series of community rules that ensure the proper functioning of the same. This is because, although the members of the community are presupposed benevolent, in a community of agents, we have to take into account individual goals and deviations that may occur on the overall goals of the community.

Something remarkable is that, the Intelligent Communities must possess the ability to function in a distributed way at the community level and the services they provide should not be embedded in the members of it, but is necessary they are implemented in a distributed way, with the propose to free the community structure, and the member agents, of computationally load.

Specialization is a feature that gives to an individual belonging to a group, a peculiarity within that group. This specialization is observed within organizations as a means to achieve greater efficiency in the objectives pursued by it. The specialization, in the field of agents has been discussed in depth by [17][18][19] among others. This specialization aims to provide a multi-agent system more efficient, so that there is an improvement in the achievement of its overall objectives. One aspect to consider is the type of specialization needed that adapts to a group or organization. This typology must implement improvements in processes and objectives.

We define the specialization of tasks as the number of different tasks to develop in a working place, and how often they are repeated, and the power of decision exercised by the working place over the design of itself. The specialization of tasks, in turn, is divided into horizontal and vertical specialization of tasks [20]. The relationship between these two types of specialization is too narrow, since, a job in which there is a high horizontal specialization also requires a high vertical specialization, because the person performing a given task can lose the overview of objectives, and therefore it is necessary that another person, who has the vision, plan, organize and control the work [20].

The organizative specialization approach can be seen as a unit, across enterprise networks. In literature, the concept of enterprise network is found in the works of authors cited in [22], for which, a set of groups, institutions or organizations can interact with each other, getting favorable results, as much individual units in the network, as a whole. Thus, a corporate network can be considered as an association of companies working together so that there is a specialized complementation between them, with the aim of resolving situations which could not resolve individually, in the most efficient way possible.

3 SCODA

SCODA (Distributed and Specialized Agent Communities) [21] is a new architecture that focuses on the development of multi-agent systems. It is based on five principles: standard, specialization, ease of implementation, reuse and distributed computing. SCODA integrates one or more Intelligent Communities provided specialization, which we call Specialized Intelligent Communities (CIE) [21]. These CIE have the ability to function as an multi-agent system, independent and specialized, where the services offered are implemented in a distributed way, so as to pursue a global goal. This structure allows to different CIE collaborate in achieving objectives, that individually they cannot reach. This philosophy is based on the *"Enterprise Networks"* [22] through which a set of groups, institutions, or organizations interact, to obtain good results as much individual units in the network as a whole. It is precisely this approach that makes SCODA is based on the concept of Intelligent Community, as an organizational unit, with the features needed to carry out their development in these terms.

Table 1 Principles of SCODA

Principle	Description
Standard	Different Specialized Smart Communities have the same structure, that is independent of the purpose to be achieved in the multi-agent system to implement
Specialization	This principle is based on human specialization, and in the Enterprise Networks, so that there is cooperation between Specialized Intelligent Communities.
Facility of Implementation	The multi-agent systems, within the context of architecture SCODA, must be easy to implement. This is achieved because the structure of the Specialized Intelligent Communities is standard, being the services they offer, which they have to be programmed.
Reusing	Having in account that, within SCODA, each Specialized Intelligent Community is adopted, as an independent multi-agent system, the reuse of these Intelligent Communities Specialized has to be viable in any SCODA based development.
Distributed Computation	The required services are not provided directly by agents of Specialized Intelligent Communities. They are running in a distributed way so that the computational load associated with agents, decreases, and structure of the architecture has no variations.

SCODA is an architecture that can be implemented on any platform for multi-agent systems that support BDI agents [23][24]. To carry out the development and implementation of this architecture is selected JADEX [25][26] because it is considered a reasoning engine and can be executed independently. SCODA architecture is based on five principles, through which seek greater efficiency of multi-agent systems developed with it.

SCODA is structured in a modular way, so that the agents, who compose it, can manage and coordinate the architecture and its functionality. It defines six basic modules: External Applications, Communications Protocol, Control, platform of agents, Specialized Intelligent Communities, and Services of the Communities. External applications are programs and users that use SCODA, requesting the services offered. The communication protocol is responsible for meeting the demands of external applications and request a response to the Specialized Intelligent Communities. The Control Module realises the follow-up of the coordination and functionality of the architecture, taking in account a fault tolerance policy in operation. The Specialized Intelligent Communities are the core of SCODA. Through by them are effective requests and responses, deliberately and of optimized form. The Community Services that running in a distributed way, is where find, the ability to processing each Specialized Intelligent Community. Finally, the agent platform, represents the environment where SCODA runs, and is composed of the agents that make up the architecture, and of the Specialized Intelligent Communities.

The agents that compose SCODA are deliberative agents BDI [23][24][27][28] and the services offered by CIE are managed by this type of agents. Another feature that is applied to all agents that make SCODA, is that being deliberative BDI agents, they can make use of reasoning mechanisms and learning techniques to perform the manage functionalities and coordination of them, depending on the particularities of the context where they run.

4 Results

To evaluate the system, we have the support of a company named Bahia Príncipe Congelados, which is specializing in the sale and distribution of frozen products. The services we have developed on SCODA, have been sales forecasting, inventory management and route optimization. The assessment raises a number of issues that affect both the architecture and in the result of the execution of the services associated with each community and its integration from SCODA. The system evaluation was performed for five months between 01/01/2012 and 31/05/2012, during which the deployed multi-agent system has worked as proposed in this section, so that allowed us to gather the information necessary to perform an analysis of their performance and draw a number of conclusions.

Over these five months, the system shows an optimal performance, in terms of integration, since it has been found that there have been only sporadic errors that correspond to failures in services of the communities. These failures have been produced by unexpected errors in the server hosting the services of CIE, because taking account on the principle of distributed computing, services are hosted on a different server to the CIE.

The specialization of the CIE, allows them to be functional individually, and by themselves solve simple problems, for which, its associated services are qualified. Also, the union of the specialization of each one of CIE, emulating an enterprise network, will obtain the benefit of solving more complex problems so that each contributes its capabilities, being SCODA binding common framework and orchestrating them. This translates into the ability of the CIE implemented, to cooperate together and produce the result of the target system, taking account that these implementations can be used in other systems, independently of their targets.

In Table 2 are show the results of applying route optimization over SCODA. The benefits obtained are important, due to daily fuel saving, that multiplied by a large fleet of trucks, has a positive effect on the company. The subsystem of inventories management, and the prediction subsystem, they do not obtain a results as good as routes optimization.

Table 2 Results on Route Optimization

Route	Actual Km	Optimized Km	Improved Km
Route A	390,4	383,9	6,5
Route B	329,8	283,7	46,1
Route C	147,9	113,8	34,1
Route D	89,8	89,8	0
Route E	356,1	279,2	76,9

5 Conclusions

The concept of community as a type of organizational unit joined to the concept of specialization, applied to multi-agent systems, have allowed to develop an architecture, named (SCODA), for developing, distributed multi-agent systems, regardless of the end that they pursue. The implementation of a system that can resolve logistics problems, has allowed us to test the operation of the architecture in a live environment, yielding promising results as much in, its integration as in terms of solving the given problems.

From these results it intends to continue working in this direction, and getting future developments in other areas.

Acknowledgments. This work has been partially supported by the MICINN project TIN 2009-13839-C03-03.

References

1. Rashid, A., Moreira, A., Araujo, J.: Modularization and composition of aspectual requirements. In: 2nd International Conference on Aspect-Oriented Software Development, pp. 11–20 (2003)
2. Kiczales, G., Lamping, J., Mendhekar, A., Maeda, C., Lopes, C., Loingtier, J., Irwin, J.: Aspect-Oriented Programming. In: Akşit, M., Matsuoka, S. (eds.) ECOOP 1997. LNCS, vol. 1241, pp. 220–242. Springer, Heidelberg (1997)
3. Zambonelli, F., Jennings, N.R., Wooldridge, M.: Organizational abstractions for the analysis and design of multi-agent systems. In: Ciancarini, P., Wooldridge, M.J. (eds.) AOSE 2000. LNCS, vol. 1957, pp. 235–251. Springer, Heidelberg (2001)
4. Ferber, J., Gutknecht, O., Michel, F.: From agents to organizations: An organizational view of multi-agent systems. In: Giorgini, P., Müller, J.P., Odell, J.J. (eds.) AOSE 2003. LNCS, vol. 2935, pp. 214–230. Springer, Heidelberg (2004)
5. Hübner, J.F., Sichman, J.S., Boissier, O.: S-Moise $^+$: A middleware for developing organised multi-agent systems. In: Boissier, O., Padget, J., Dignum, V., Lindemann, G., Matson, E., Ossowski, S., Sichman, J.S., Vázquez-Salceda, J. (eds.) ANIREM 2005 and OOOP 2005. LNCS (LNAI), vol. 3913, pp. 64–78. Springer, Heidelberg (2006)
6. Zambonelli, F., Jennings, N., Wooldridge, M.: Developing Multiagent Systems: The Gaia Methodology. ACM Transactions on Software Engineering and Methodology 12, 317–370 (2003)
7. Pavón, J., Gómez-Sanz, J.: Agent Oriented Software Engineering with INGENIAS. In: Mařík, V., Pěchouček, M., Müller, J. (eds.) CEEMAS 2003. LNCS (LNAI), vol. 2691, pp. 394–403. Springer, Heidelberg (2003)
8. Pavón, J., Gómez-Sanz, J.J., Fuentes, R.: The INGENIAS Methodology and Tools. In: Henderson-Sellers, B., Giorgini, P. (eds.) Agent Oriented Methodologies, pp. 236–276. IDEA Group Publishing (2005)
9. Dignum, V., Dignum, F.: A landscape of agent systems for the real world. Technical Report 44-CS-2006-061. Institute of Information and Computing Sciences, Utrecht University (2006)
10. Corchado, J.M.: Agentes Software y Sistemas Multiagente. Pearson Education (2005), Pérez de la Cruz, J.L. (ed.)
11. Snyder, R., MacKenzie, D.: Cougaar Agent Communities. In: Proceedings of the 1st Open Cougaar Conference, New York City, pp. 143–148 (2004)
12. Nicklisch, H.: Cuestiones fundamentales de la Economía de Empresa, Stuttgart (1928)
13. Nicklisch, H.: La economía de empresa, Stuttgart (1932)
14. Larsen, B.: German organization and leadership theory-stable trends and flexible adaptation. Scandinavian Journal of Management 19(2003), 103–133 (2000)
15. Glinton, R., Paruchuri, P., Scerri, P., Sycara, K.: Self-Organized Criticality of Belief Propagation in Large Heterogeneous Teams. In: Hirsch, M.J., Pardalos, P.M., Murphey, R. (eds.) Dynamics of Information Systems: Theory and Applications. Springer, Berlin (2010)
16. Parachuri, P., Glinton, R., Sycara, K., Scerri, P.: Effect of humans on belief propagation in large heterogeneous teams. In: Hirsch, M.J., Pardalos, P.M., Murphey, R. (eds.) Dynamics of Information Systems: Theory and Applications. Springer, Berlin (2010)

17. Theraulaz, G., Gervet, J., Semenoff, S.: Social regulation of foraging activities in polistes dominulus christ: a systemic approach to behavioural organization. Behaviour 116(1), 292–320 (1991)
18. Chai, L., Chen, J., Han, Z., Di, Z., Fan, Y.: Emergence of Specialization from Global Optimizing Evolution in a Multi-agent System. In: Shi, Y., van Albada, G.D., Dongarra, J., Sloot, P.M.A. (eds.) ICCS 2007, Part IV. LNCS, vol. 4490, pp. 98–105. Springer, Heidelberg (2007)
19. Okamoto, S., Scerri, P., Sycara, K.: The Impact of Vertical Specialization on Hierarchical Multi-Agent Systems. In: Proceedings of the Twenty-Third AAAI Conference on Artificial Intelligence (2008)
20. Mintzberg, H.: El diseño de las organizaciones eficientes. Ed. El Ateneo. Argentina (1989)
21. Román, J.A., Tapia, D.I., Corchado, J.M.: SCODA para el Desarrollo de Sistemas Multiagente. Revista Ibérica de Sistemas y Tecnologías de Información 8, 25–38 (2011)
22. Becerra, F.: Las redes empresariales y la dinámica de la empresa: aproximación teórica. INNOVAR. Revista de Ciencias Administrativas y Sociales (en línea), 18 (2008), Disponible en `http://redalyc.uaemex.mx/src/inicio/ArtPdfRed.jsp?iCve=81803203`
23. Bratman, M.E., Israel, D., Pollack, M.: Plans and resource-bounded practical reasoning. Computational Intelligence 4, 349–355 (1988)
24. Rao, A.S., Georgeff, M.P.: BDI Agents from Theory to Practice. In: Proceedings of the First International Conference on Multi-Agents Systems (ICMAS 1995), pp. 312–319 (1995)
25. Pokahr, A., Braubach, L., Lamersdorf, W.: Jadex: Implementing a BDI-Infrastructure for JADE Agents. In EXP - in search of innovation (Special Issue on JADE), 76–85 (2003)
26. Pokahr, A., Braubach, L., Walczak, A., Lamersdorf, W.: Jadex - Engineering Goal-Oriented Agents. In: Developing Multi-Agent Systems with JADE, pp. 254–258. Wiley & Sons (2007)
27. Koster, A., Schorlemmer, M., Sabater, J.: Opening the black box of trust: reasoning about trust models in a BDI agent. Journal of Logic and Computation (2012)
28. van Oijen, J., Dignum, F.: Towards a Design Approach for Integrating BDI Agents in Virtual Environments. In: Vilhjálmsson, H.H., Kopp, S., Marsella, S., Thórisson, K.R. (eds.) IVA 2011. LNCS, vol. 6895, pp. 462–463. Springer, Heidelberg (2011)
29. Posland, S.: Specifying protocols for multi-agent systems interaction. ACM Trans. Auton. Adapt. Systems 2(4), 15 (2007)

A Gateway Protocol Based on FIPA-ACL for the New Agent Platform PANGEA

Alejandro Sánchez, Gabriel Villarrubia, Carolina Zato,
Sara Rodríguez, and Pablo Chamoso

Abstract. Communication is one of the cornerstones of the intelligent agents paradigm. There are different forms of communication between agents, just as there are many platforms for creating them. However, one of the problems we encountered when using the agent paradigm is the actual communication between platforms. That is, to have a gateway of communication between different types of agents regardless of the platform has been used to create them. To this end, a new way of communication between PANGEA and other multiagent platforms is presented in this paper. In this communication process the FIPA-ACL standards are used.

Keywords: PANGEA, Multi-agent systems, Gateway, Communication agents, ACL messages, FIPA-ACL.

1 Introduction

Currently, multi-agent systems (MAS) have become an efficient mechanism for the development of tools and distributed applications with strong needs of scalability, interaction and of course, gifted with intelligence at different levels. The main problem is that each tool is built on a different agent platform and in most cases it is difficult to communicate with each other platforms despite the FIPA [7] standards. This drawback forces to build middlewares or frameworks to enable this interaction between systems running on different platforms. This paper presents the development of a gateway to external communication with the platform called PANGEA [28] using FIPA and IRC standards. This gateway allows external agents to get some result from

Alejandro Sánchez · Gabriel Villarrubia · Carolina Zato ·
Sara Rodríguez · Pablo Chamoso
Department of Science Computing and Automation,
University of Salamanca, Plaza de la Merced, S/N,
37008, Salamanca, Spain
e-mail: {asanchezyu,gvg,carol_zato,srg,chamoso}@usal.es

J.B. Pérez et al. (Eds.): *Trends in Prac. Appl. of Agents & Multiagent Syst.*, AISC 221, pp. 41–51.
DOI: 10.1007/978-3-319-00563-8_6

PANGEA. Next section introduces the problem and explains why there is a need of agents communication between different platforms. The following section describes the main features and capabilities of the system. Finally we will present the results and conclusions, including the future work.

2 Background

When discussing MAS, the idea of a single agent is expanded to include an infrastructure for interaction and communication. Ideally, MAS include the following characteristics [13]: (i) they are typically open with a non-centralized design; (ii) they contain agents that are autonomous, heterogeneous and distributed each with its own personality (cooperative, selfish, honest, etc.). They provide an infrastructure specifically for communication and interaction protocols. Multi-agent systems allow the participation of agents within different architectures and even different languages [27] [5]. The development of open MAS is still a recent field of the multi-agent systems paradigm and its development will allow applying the agent technology in new and more complex application domains. Open MAS should allow the participation of heterogeneous agents with different architectures and even different languages [27]. However, currently, most of the multi-agent platforms (JADE [2],JACK [11], Jadex [13], Jason [3], S-MOISE+ [12], J-MOISE [14], Janus [9], MadKit [10], Cartago [24], NetLogo [26], SeSAm [19], Magique [25], MALEVA [4], Malaca [1]), are not able to communicate with agents of other platforms different than their own. Nowadays, the agent communication standard is characterized by ACL standard [6] (Agent Communication Language) of the Foundation for Intelligent Physical Agents (FIPA). This language marks a type of message, which contains the necessary fields for a good communication between agents [8]. The message structure has the following fields:

- Participants in communication: sender, receiver, reply-to.
- Performative.
- Content.
- Description of content: Language, encoding, ontology.
- Control of conversation: Protocol, conversation-id, reply-with, in-reply-to, reply-by.

All these fields have been defined by FIPA for a complete agent communication and any parameters will not be missed for the coherence of the message. Almost all existing multi-agent systems are using the standard FIPA-ACL for communication between agents, such as JADE [2],Jadex [23], JACK [11], Janus [9], MadKit [10], and NetLogo [26]. There are previous studies that try to create middleware for heterogeneous communication such as [21]. Other studies focus on a gateway platform with web services [22](not a middleware). In our case, the implementation is to join several platforms with PANGEA [28]. PANGEA allows us to create virtual organizations of agents, and therefore, agents, regardless of platform and language. Since FIPA is the standard, the work presented in this paper, it also allows interaction between players who use this means of communication used in many other agents platforms.

3 Pangea Overview

PANGEA is a service oriented platform that allows the implemented open MAS to take maximum advantage of the distribution of resources. To this end, all services are implemented as Web Services [29]. Due to its service orientation, different tools modeled with agents that consume Web services can be integrated and operated from the platform, regardless of their physical location or implementation. This makes it possible for the platform to include both a service provider agent and a consumer agent, thus emulating client-server architecture. The provider agent (a general agent that provides a service) knows how to contact the web service, while the remaining agents know how to contact with the provider agent due to their communication with the ServiceAgent, which contains information about services. Once the client agents request has been received, the provider agent extracts the required parameters and establishes contact. Once received, the results are sent to the client agent. Using Web Services also allows the platform to introduce the SOA (Service-oriented Architecture) into MAS systems. SOA is an architectural style for building applications that use services available in a network such as the web. It promotes loose coupling between software components so that they can be reused. Applications in SOA are built based on services.

Using PANGEA, the platform will automatically launch the following agents:

- OrganizationManager: this agent is responsible for the actual management of organizations and suborganizations. It is responsible for verifying the entry and exit of agents, and for assigning roles. To carry out these tasks, it works with the OrganizationAgent, which is a specialized version of this agent.
- InformationAgent: this agent is responsible for accessing the database containing all pertinent system information.
- ServiceAgent: this agent is responsible for recording and controlling the operation of services offered by the agents.
- NormAgent: this agent ensures compliance with all the refined norms in the organization.
- CommunicationAgent: this agent is responsible for controlling communication among agents, and for recording the interaction between agents and or-ganizations.
- Sniffer: manages the message history and filters information by controlling communication initiated by queries.

The platform agents are implemented with Java, while the rest of the agents may be implemented in other programming languages. Communication between agents has been based on the IRC protocol. The IRC protocol was used to implement communication. Internet Relay Chat (IRC) is a real time internet protocol for sim-ultaneous text messaging or conferencing. This protocol is regulated by 5 standards: RFC1459 [20], RFC2810 [15], RFC2811 [16], RFC2812 [17] and RFC2813 [18]. It is designed primarily for group conversations in discussion forums and channel calls, but also allows private messaging for one on one communications, and data transfers, including file exchanges [20]. The protocol in the OSI model I located on the application layer and uses TCP or alternatively TLS [16]. An IRC server can connect with other IRC servers to expand the user network. Users access the IRC networks by connecting a client to a server. There have been many imple-mentations of clients, including mIRC or XChat. The original protocol is based on flat text (although it was subsequently expanded), and uses TCP port 6667 as its primary port, or other nearby ports (for example TCP ports 6660-6669, 7000) [17]. The standard structure for an IRC server network is a tree configuration. The messages are routed only through those nodes that are strictly necessary; however, the network status is sent to all servers. When a message must be sent to multiple recipients, it is sent similarly to a multidiffusion; that is, each message is sent to a network link only once [15]. This is a strong point in its favor compared to the no-multicast protocols such as SimpleMail Transfer Protocol (SMTP) or the Extensible Messaging and Presence Protocol (XMPP). One of the most important features that characterizes the platform is the use of the IRC protocol for communication among agents. This allows for the use of a protocol that is easy to implement, flexible and robust. The open standard protocol enables its continuous evolution. There are also IRC clients for all operating systems, including mobile devices.

3.1 Gateway Pangea-ACL Messages

The gateway arises from the need to communicate external agents with PANGEA multiagent system. The external agent must send an FIPA-ACL object to the IP ad-dress of the PANGEA MAS, through the 6668 port. The object must have all the necessary fields for a good communication. This is the case of the ontology, content, sender, etc. The responsible for making inside-outside communication is the ACLAgent, which is deployed in PANGEA. This special agent is the responsible of converting FIPA-ACL messages in PANGEA messages. In addition to this, it also makes the demanded operations and it must return the results by an ACL object. At first, it takes cares of the operations based on the request, inform, subscription, and contract-net protocols. The subsections below shows these operations based on the protocols. We use JAVA introspection to differentiate the protocols mentioned before. The ACLAgent manages the input object via Java introspection as follows:

Request Protocol

This kind of operation is used in order to get a result of a known web service which it is offered in PANGEA platform. In Figure 1 it is possible to observe the workflow that is followed in this protocol. First of all, the external agent sends the object to the ACLAgent, The ACLAgent sends a message to OrganizationManager to get a service result. The OrganizationManager talks with the InformationAgent if there is a service with a specific name. If there is not a service with the same description, the OrganizationManager ask to execute the service and get the result. Finally ACLAgent sends the response to the external agent.

if *object.protocol equals REQUEST* **then**
  send request to ServiceAgent;
  receive Response;
  send Response to external agent;
end

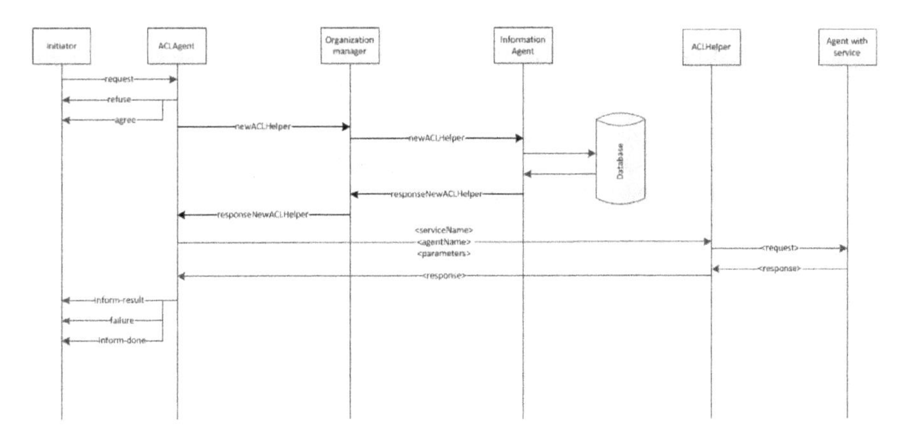

Fig. 1 Request workflow

Inform Protocol

The inform protocol is used to inform PANGEA that a new service is offered. So, the external agent sends a message with the content right format and it will receive an answer (refuse, failure, inform-done). In Figure 2, the inform workflow is shown. The external agent wants to register a new service to Pangea MAS. It sends an object to the ACLAgent with the service description. The ACLAgent sends it to the OrganizationManager and it registers the service. Finally, ACLAgent sends a response to the external agent.

if *object.protocol equals INFORM* **then**
  send new Service to OrganizationManager;
  send Response to external agent;
end

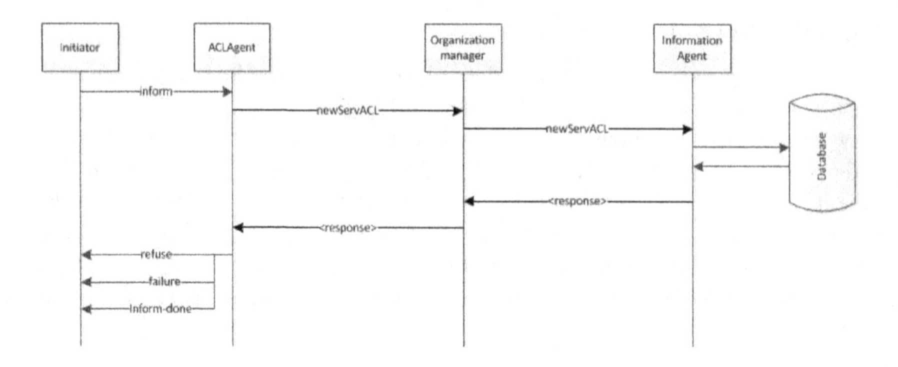

Fig. 2 Inform workflow

Subscription Protocol

External agent makes the request of a subscription. The ACLAgent will return it every news or modifications made in that subscription. In Figure 3 we can see the subscribe workflow.

if *object.protocol equals SUBSCRIPTION* **then**
  create subscriptionHelperAgent;
  if *message received from subscription is TRUE* **then**
    send Message to external agent;
  end
end

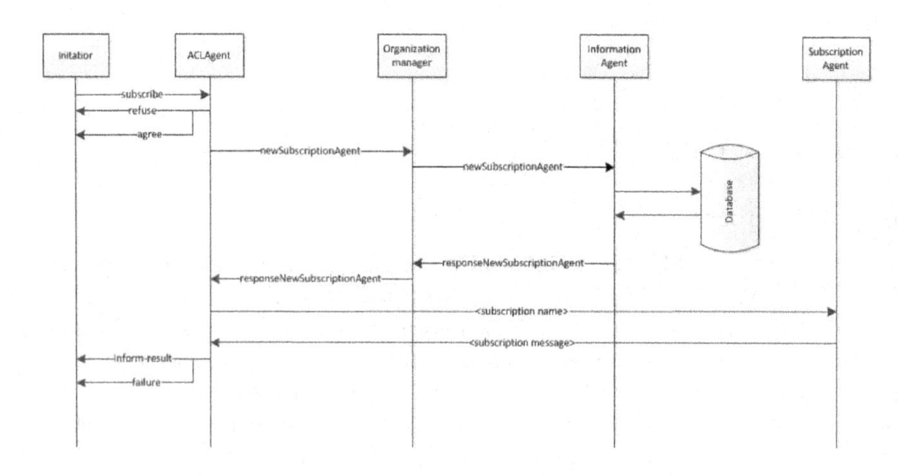

Fig. 3 Subscribe workflow

Contract-Net Protocol

In this protocol, the external agent can search, find and execute a specific service. For this purpose, sending multiple messages is needed. Firstly, the name of the service which will be executed, must be known, so, a CFP message (Call for proposals) is sent with a little description of the service needed in the content field. Several answers will be sent to the external agent. The agent will choose one of them, and the chosen one will be executed. In Figure 4, we can see the contract net protocol workflow.

if *object.protocol equals CONTRACT-NET* **then**
  send request to ServiceAgent;
  receive service names;
  send response to external agent;
  if *response from outside* **then**
    send request to ServiceAgent;
    receive Response;
    send Response to external agent;
  end
end

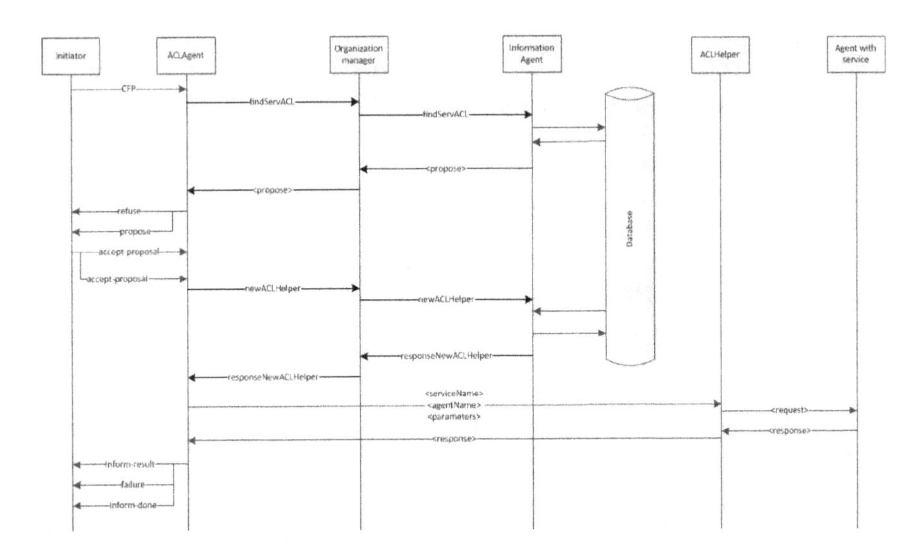

Fig. 4 Contract-net workflow

In JADE, sending a message is instant, the programmer gives the order, and the message is sent ipso facto. The ACLAgent does not it in this way. It has a messaging queue, so that when a programmer gives the order to send a message, the message is automatically inserted in the scheduled queue and manages the input object via Java introspection. In Figure 5, a comparative picture between JADE and the changes made in the gateway with PANGEA is shown.

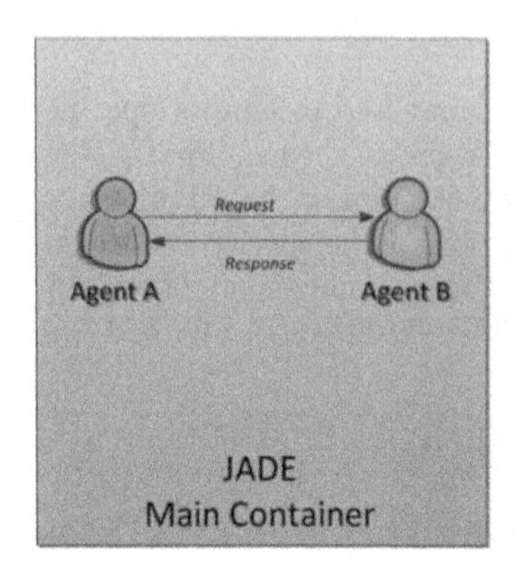

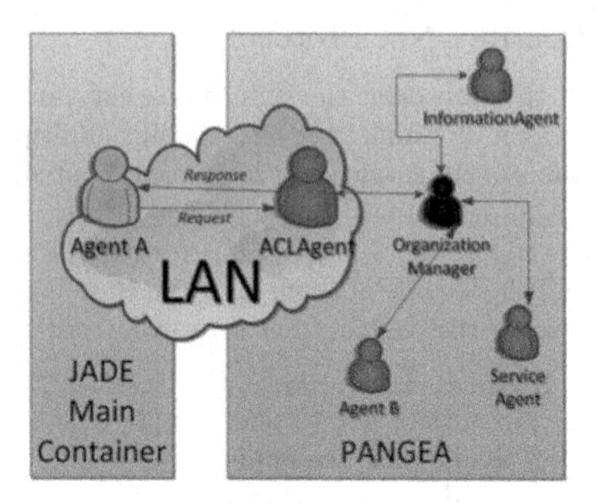

Fig. 5 (a) Sending a message in JADE. (b) Sending a message incorporating PANGEA

4 Case Study

The case of study consists on the communication between the PANGEA platform and some agents developed in the JADE platform. Three computers are involved in this test; they are connected to the same LAN. In one of these computers, PANGEA has been deployed with all the control agents, including the ACLAgent, which is the responsible for the external communication. In a second machine, we have other PANGEA agents which offer services to PANGEA multi-agent system. One of these agents offers solutions for basic arithmetical operations (addition, multiplication, division, and substraction). The other agent offers news from a RSS channel. These are the agents that are asked for the services. The external agents are deployed in the last computer. These agents are developed in JADE platform. The only thing we have to do in these agents, for being able to send messages to PANGEA platform, is to send a jade.lang.acl.ACLMessage object by 6668 port. Tests have consisted in the execution of the four different options we can make. We can divide the test in two groups:

- Simple workflow tests: These the tests of the protocols INFORM and REQUEST are. In these tests, the workflow consists in question-answer. There are one only input, and one only output. Both tests have been satis-factory, because, correct outputs have been obtained.
- Complex workflow tests: These the tests of the protocols SUBSCRIBE and CONTRACT-NET are. In the first case, it is composed by one input and several outputs (one per event done in the subscription). In the other hand, the other protocol consists on a communication between both parts, in which there are several input and output messages. Both parts must have to be listening for possible answers of the other agent. The obtained results of these tests have been satisfying, especially the part of the gateway,

which has work perfectly. On the other hand, if the external agent does not control the flow of the messages, some of these will be lost and the communication will be incomplete.

5 Results and Conclusions

With the development of this gateway, the communication between different architectures has been achieved. The gateway allows several multi-agent systems work jointly for a common target, and share all its knowledge. After the made tests during its development, obtained data has been significantly positive, because communication between both parts is very fluid and fast, so you hardly notice they are working on different machines. This is in that way because of the low weight of the packet sent through the LAN and the PANGEAs message management. Thanks to the use of the gateway, new communication ways are opened. External agents will benefit from the services of the PANGEA platform, and vice versa. PANGEA agents will ask for some tasks to the outside of the platform. The implementation of the rest of the FIPA-ACL communication protocols have been proposed as future work. The rests of the protocols will make a complete com-munication, and all kind of queries would be made to PANGEA platform. It is also though to port and develop this idea in mobile environments.

Acknowledgements. This research has been supported by the project OVAMAH (TIN2009-13839-C03-03) funded by the Spanish Ministry of Science and Innovation.

References

1. Amor, M., Fuentes, L.: Malaca: A Component and Aspect-Oriented Agent Architecture. Information and Software Technology, 1052–1065 (2009)
2. Bellifemine, F., Poggi, A., Rimassa, G.: JADE A FIPA-compliant agent framework. In: Proceedings of the Practical Applications of Intelligent Agents (1999)
3. Bordini, R.H., Hübner, J.F., Wooldridge, M.: Programming multi-agent systems in Agent Speak using Jason (2008)
4. Briot, J.-P., Meurisse, T., Peschanski, F.: Architectural Design of Component-Based Agents. In: Bordini, R.H., Dastani, M., Dix, J., El Fallah Seghrouchni, A. (eds.) ProMAS 2006. LNCS (LNAI), vol. 4411, pp. 71–90. Springer, Heidelberg (2007)
5. Corchado, E., Pellicer, M.A., Borrajo, M.L.: A MLHL Based Method to an Agent-Based Architecture. International Journal of Computer Mathematics 86(10 & 11), 1760–1768 (2008); ISI JCR Impact Factor: 0.423
6. FIPA ACL Message Structure Specification, `http://www.fipa.org/specs/fipa00061/index.html`
7. FIPA, The Foundation for Intelligent Physical Agents, `http://www.fipa.org/`
8. Foundation for Intelligent Physical Agents - FIPA ACL Message Structure Specification (2002), `http://www.fipa.org/specs/fipa00061/SC00061G.pdf`

9. Gaud, N., Galland, S., Hilaire, V., Koukam, A.: An Organisational Platform for Holonic and Multiagent Systems. In: Hindriks, K.V., Pokahr, A., Sardina, S. (eds.) ProMAS 2008. LNCS, vol. 5442, pp. 104–119. Springer, Heidelberg (2009)
10. Gutknecht, O., Ferber, J.: The MADKIT Agent Platform Architecture. In: Wagner, T.A., Rana, O.F. (eds.) AA-WS 2000. LNCS (LNAI), vol. 1887, pp. 48–55. Springer, Heidelberg (2001)
11. Howden, N., Rönnquist, R., Hodgson, A., Lucas, A.: JACK Intelligent Agents - Summary of an Agent Infrastructure. In: 5th International Conference on Autonomous Agents (2001)
12. Hübner, J.F., Sichman, J., Boissier, O.: Developing organised multiagent systems using the MOISE+ model: programming issues at the system and agent levels. International Journal of Agent-Oriented Software Engineering, 370–395 (2007)
13. Huhns, M., Stephens, L.: Multiagent Systems and Societies of Agents. In: Weiss, G. (ed.) Multi-agent Systems: a Modern Approach to Distributed Artificial Intelligence. MIT (1999)
14. Hübner, J.F., Sichman, J., Boissier, O.: Developing organised multiagent systems using the MOISE+ model: programming issues at the system and agent levels. International Journal of Agent-Oriented Software Engineering 1(3), 370–395 (2007)
15. Kalt, C.: Internet Relay Chat: Architecture, RFC 2811 (April 2000)
16. Kalt, C.: Internet Relay Chat: Channel Management, RFC 2811 (April 2000)
17. Kalt, C.: Internet Relay Chat: Client Protocol, RFC 2812 (April 2000)
18. Kalt, C.: Internet Relay Chat: Server Protocol, RFC 2813 (April 2000)
19. Klügl, F., Herrler, R., Oechslein, C.: From Simulated to Real Environments: How to Use SeSAm for Software Development. In: Schillo, M., Klusch, M., Müller, J., Tianfield, H. (eds.) MATES 2003. LNCS (LNAI), vol. 2831, pp. 13–24. Springer, Heidelberg (2003)
20. Oikarinen, J., Reed, D.: Internet Relay Chat Protocol, RFC 1459 (May 1993)
21. Omair Shafiq, M., Ali, A., Farooq Ahmad, H., Suguri, H.: AgentWeb Gateway - a middleware for dynamic integration of multi agent system and Web services framework. In: 14th IEEE International Workshops on Enabling Technologies: Infrastructure for Collaborative Enterprise, June 13-15, pp. 267–268 (2005)
22. Omair Shafiq, M., Ding, Y., Fensel, D.: Bridging Multi Agent Systems and Web Services: towards interoperability between Software Agents and Semantic Web Services. In: 10th IEEE International Conference on Enterprise Distributed Object Computing, EDOC 2006, pp. 85–96 (October 2006)
23. Pokahr, A., Braubach, L., Lamersdorf, W.: Jadex: A BDI Reasoning Engine. In: Multi-Agent Programming: Languages, Platforms and Applications, pp. 149–174 (2005)
24. Ricci, A., Viroli, M., Omicini, A.: CArtAgO: A Framework for Prototyping Artifact-Based Environments in MAS. In: Weyns, D., Van Dyke Parunak, H., Michel, F. (eds.) E4MAS 2006. LNCS (LNAI), vol. 4389, pp. 67–86. Springer, Heidelberg (2007)
25. Routier, J.C., Mathieu, P., Secq, Y.: Dynamic Skills Learning: A Support to Agent Evolution. In: Proceedings of the Artificial Intelligence and the Simulation of Behaviour, Symposium on Adaptive Agents and Multi-agent Systems, AISB 2001 (2001)

26. Wilensky, U.: NetLogo itself. Center for Connected Learning and Computer-Based Modeling, Northwestern University, Evanston, IL (1999), http://ccl.northwestern.edu/netlogo/
27. Zambonelli, F., Jennings, N.R., Wooldridge, M.: Developing Multiagent Systems: The Gaia Methodology. ACM Transactions on Software Engineering and Methodology 12, 317–370 (2003)
28. Zato, C., Villarrubia, G., Sánchez, A., Barri, I., Rubión, E., Fernández, A., Rebate, C., Cabo, J.A., Álamos, T., Sanz, J., Seco, J., Bajo, J., Corchado, J.M.: PANGEA – Platform for Automatic coNstruction of orGanizations of intElligent Agents. In: Omatu, S., Paz Santana, J.F., González, S.R., Molina, J.M., Bernardos, A.M., Rodríguez, J.M.C. (eds.) Distributed Computing and Artificial Intelligence. AISC, vol. 151, pp. 229–239. Springer, Heidelberg (2012)
29. Zato, C., Sanchez, A., Villarrubia, G., Rodriguez, S., Corchado, J.M., Bajo, J.: Platform for building large-scale agent-based systems. In: 2012 IEEE Conference on Evolving and Adaptive Intelligent Systems (EAIS), May 17-18, pp. 69–73 (2012)

Applying Classifiers in Indoor Location System

Gabriel Villarubia, Francisco Rubio, Juan F. De Paz,
Javier Bajo, and Carolina Zato

Abstract. Research in indoor location has acquired a growing importance during the recent years. The main objective is to obtain functional systems able of providing the most precise location, identification and guidance in real time. Currently, none of the existing indoor solutions have obtained location or navigation results as precise as the ones provided by the analog systems used outdoor, such as GPS. This paper presents an indoor location system based on Wi-Fi technology which, from the use of intensity maps and classifiers, allows effective and precise indoor location.

Keywords: indoor location system, classifiers, Wi-Fi.

1 Introduction

Nowadays, indoor location is one of the most promising research fields in mobile computing [7], and it is possible to find several studies. To obtain precise information about the location of objects or people inside building can provide very useful information to develop services. Some of the most relevant of these services are those oriented to identification-based access control, location-based security, location-aware computing, etc.. These services are deployed in indoor environments, such as hospitals, factories, shopping malls, or even as complementary systems to GPS (Global Positioning System). The main challenge in current indoor location systems is to obtain precise measures with a reasonable cost in infrastructure.

Gabriel Villarubia · Francisco Rubio · Juan F. De Paz · Carolina Zato
Department of Computer Science and Automation, University of Salamanca,
Plaza de la Merced, s/n, 37008, Salamanca, Spain
e-mail: {gvg,paco_rg,fcofds,carol_zato}@usal.es

Javier Bajo
Departamento de Inteligencia Artificial, Facultad de Informática,
Universidad Politécnica de Madrid, Madrid, Spain
e-mail: jbajo@fi.upm.es

J.B. Pérez et al. (Eds.): *Trends in Prac. Appl. of Agents & Multiagent Syst.*, AISC 221, pp. 53–58.
DOI: 10.1007/978-3-319-00563-8_7

The main reason to explain why it is not possible to find effective indoor solutions is due to technical and financial reasons. If we take into consideration a GPS system, we simply need a physical device that connects to a finite number of satellites in open spaces. However, in an open space, it is required to make use of a technological infrastructure with a considerable number of fixed devices that are used as beacons or readers, and notably improve the cost of the system. Amongst the technologies that are currently used most in the development of Real-Time Location Systems are, RFID (Radio Frequency IDentification) [8] [10], Wi-Fi y ZigBee [8] [9]. Nowadays, some of the existing indoor location engines make use of 802.11 technology also known as Wi-Fi, and have acquired a growing importance due to the reduced and reasonable cost of these systems compared to the alternatives. Moreover, Wi-Fi provides a good number of advantages: easy deployment and is integrated in most of the current electronic devices, wide range of existing networks installed in several buildings and locations.

This paper presents a study aimed at obtaining the design of an innovative Wi-Fi indoor location engine, able of providing location-aware Information of people or objects inside a building. The use of Wi-Fi technology for location systems is mainly based on intensity maps constructed from RSSI levels in different zones. The maps are used as a basis to obtain classifications. The classifiers use the data of the maps and the data obtained from the devices to determine the position of a person inside a building. The system presented in this paper has been adapted to be installed in resource-constrained devices. In this way it is not necessary to connect to databases to obtain information for the classification, which notable improve the performance of the system and the fault tolerance.

This article is divided as follows: section 2 describes the state of the art; section 3 presents the proposed model; section four describes the results obtained and the conclusions respectively.

2 Background

There exist three main algorithms that are used by real-time location systems to determine the location of the mobile nodes (tags): Triangulation, Fingerprinting and Multilateration [4]. Triangulation allows obtaining location coordinates by means of the calculus of the length of the sides of the triangle from the angles of the received signal in each of the antennas, which requires at least 3 reference points. Fingerprinting, also known as signpost or symbolic location, is based on the study of the characteristics of the each of the location zones, obtaining measurements of the radiofrequency characteristics and estimating the influence area of each tag. [5]. Multilateration estimates the distances between readers and tags, taking into account parameters as RSSI (Received Signal Strength Indication) or TDOA (Time Difference Of Arrival) [6], in such way that the intersection of the estimated distances from each tag to three or more fixed nodes determines the point where the tags are identified. Multilateration provides better results that triangulation when it is used outdoors, but its performance is notably reduced indoors. The reason is fluctuation of the RSSI levels in indoor environments due to the presence of different elements (people, objects, animals). Besides, multilateration is based on the estimation of distances, and it is necessary a previous estimation of the RSSI values, which is difficult because the RSSI values change constantly.

Location techniques based on triangulation and multilateration cannot be considered as efficient because the signals suffer several attenuations caused by the elements present in the rooms. In these cases, it is recommended to use heuristics for the classification process and perform training to obtain location algorithms that improve the precision. The heuristics allow the collection of possible measurements in specific positions and to use these measurements to calculate the most probable position.

3 Proposed Reasoning System

Classification techniques facilitate the association of cases to the existing groups. The behavior of these algorithms is similar to the algorithms used for clustering, but they are too much simple in most of the cases. The existing techniques can be grouped in the following categories depending on the nature of the algorithms:

- Decision rules and decision trees.
- Probabilistic models: Naive Bayes [11], Bayesian networks [1] [2] [3].
- Fuzzy logic: K-NN (K-Nearest Neighbours), NN (K-Nearest Neighbours).
- Function search: Sequential Minimal Optimization (SMO) [12].
- Artificial neural networks.

Specifically, the proposed system has integrated a bayesian network to estimate the probabilities of belonging to the points previously scanned in the intensity maps. The intensity maps obtained indoors are created using the parameters shown in Table 1. Each of the rows in Table 1 contains the information of all the Wi-Fi networks scanned in that moment and identified by a coordinate (x,y). The rows can contain more or less columns depending on the scanned intensities.

Table 1 Format of the measurements in the intensity maps

x	y	SSID	BSSID	RSSI	SSID	...

The scanned measurements, represented using the format shown in Table 1, are used to obtain the distances that are used by the bayesian network which is trained and configured for further estimations. In order to build Bayesian networks, it is first necessary to establish search mechanisms that can generate the DAG (Directed Acyclic Graph) using a set of heuristics that can reduce the number of combinations and generate the final Bayesian network. There are various Bayesian network search mechanisms, including tabu search [1], conditional independence

[2], K2 [1], HillClimber [1], TAN (Tree Argumented Naive Bayes) [3]. In this work, we have used conditional independence. This algorithm is based on the calculation of the conditional Independence test for the variables to generate a DAG that can obtain the probability estimates. If the variables being studied are independent, it will not be possible to generate a Bayesian network with good results, but in this work this is not a problem with the independence of the variables. The attribute for the classification in the bayesian network is defined as the union of x and y.

4 Results and Conclusions

To analyze the overall performance of the system, we defined a case study in the University of Salamanca. The first step was to calibrate the map for the first floor of the Physics building at the Faculty of Sciences. During the calibration process, we obtained different signal measurements in the corridors of the floor. The classifier was constructed from these data. The total number of measurements was 380, as shown in Figure 1a. The area of the floor of the building was approximately 1700m2. The floor was equipped with 4 fixed access points spread across the floor of the building. Moreover, some signals emitted by access points that went out and lit corresponding to offices and laboratories were detected. The access points were the infrastructure nodes available in the university, and it was not necessary to introduce any additional access points to reduce the system error. Figure shows a screenshot obtained from the mobile device used for the experiment.

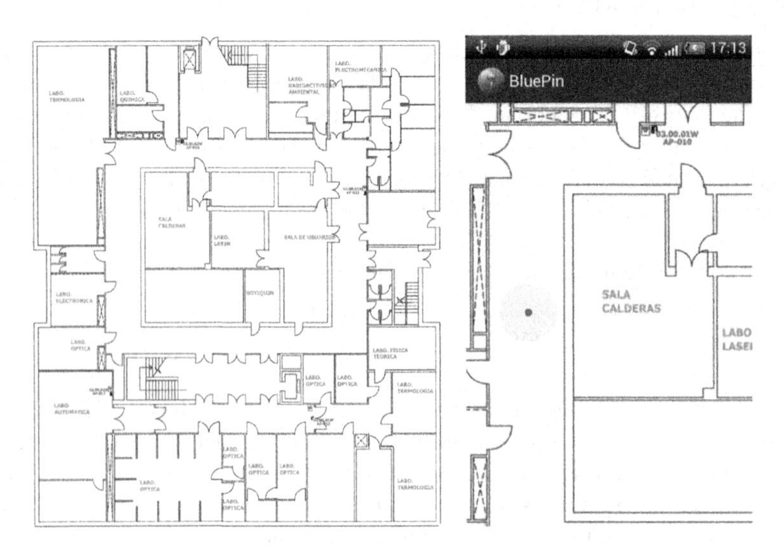

Fig. 1 A) First floor of the Physics building. B) Mobile interface of the system.

The system was evaluated obtaining measures in different points and estimating the position. We obtained 332 new measurements and estimated the corresponding positions. The results obtained are shown in Table 2. The absolute mean error obtained is shown in the first column and, as can be seen in Table 2, the bayesian network provides the best value for the error, 2,4 meters. The average time to estimate the position is 0,65 milliseconds, which is good enough to be used in a mobile device. It is necessary that the average time has been obtained in a laptop. It has not been possible to measure the time in a mobile device, but the system perfectly worked in the mobile during the experiment.

Table 2 Errors and time to estimate positions using different classifiers

Error meters	Time millisecond	Classifier
2,414	0,652632	.BayesNet
3,040	2,126316	NaiveBayes
3,040	1,957895	NaiveBayesUpdateable
4,147	37,989475	SimpleLogistic
8,890	22,660526	SMO
3,945	0,281579	Ibk
8,299	11,03421	LWL
2,857	14,047368	KStar
9,430	0,078947	AdaBoostM1
3,830	0,281579	AttributeSelectedClassifier
5,035	1,452632	Bagging
4,256	6,481579	ClassificationViaRegression
12,605	0,042105	CVParameterSelection
4,325	0,2	FilteredClassifier
4,306	8,652632	LogitBoost
12,624	7,476316	MultiClassClassifier
12,605	0,052632	MultiScheme
3,826	0,4	RandomCommittee
3,560	0,768421	RandomSubSpace
12,605	0,047368	Stacking
12,510	5,247368	DecisionTable
4,572	1,760526	JRip
13,370	0,044737	OneR
6,488	1,639474	PART
12,605	0,039474	ZeroR
9,430	0,068421	DecisionStump
7,819	0,255263	J48
14,142	90,007896	LMT
3,346	0,415789	RandomForest
2,918	0,092105	RandomTree
2,649	0,194737	REPTree

As a conclusion, it is possible to say that the work presented in this paper facilitates a new technique to locate objects and people using Wi-Fi signals in indoor environment, making use of the existing infrastructure networks. The system does not require the installation of additional access points or any other hardware to improve the system performance. Thus, the implementation cost is notably reduced compared to most of the existing systems. Our future work focuses on the incorporation of new measurements into the algorithm, such as GSM or 3G that can be combined with the Wi-Fi signals and improve the overall performance of the system.

Acknowledgments. This work has been supported by the Ministry of Economy and Competitiveness (INNPACTO) IPT-2011-0726-430000.

References

1. Bouckaert, R.R.: Bayesian Belief Networks: from Construction to Inference, Utrecht, Netherlands (1995)
2. Verma, T., Pearl, J.: An algorithm for deciding if a set of observed independencies has a causal explanation. In: Proc. of the Eighth Conference on Uncertainty in Artificial Intelligence, pp. 323–330 (1992)
3. Friedman, N., Geiger, D., Goldszmidt, M.: Bayesian Network Classifiers. Machine Learning 29, 131–163 (1997)
4. Glassner, A.: Principles of digital image synthesis. Morgan Kaufmann (1995)
5. Georgé, J.P., Gleizes, M.P., Glize, P.: Emergence of organisations, emergence of functions. In: Symposium on Adaptive Agents and Multi-Agent Systems, pp. 103–108 (2003)
6. De Paz, J.F., Rodríguez, S., Bajo, J., Corchado, J.M.: Multi-agent system for security control on industrial environments. International Transactions on System Science and Applications Journal 4(3), 222–226 (2008)
7. Chen, Y.-C., Chiang, J.-R., Chu, H.-H., Huang, P., Tsuid, A.W.: Sensor-Assisted Wi-Fi Indoor Location System for Adapting to Environmental Dynamics (2011)
8. Razavi, R.S., Perrot, J.-F., Guelfi, N.: Adaptive modeling: An approach and a method for implementing adaptive agents. In: Ishida, T., Gasser, L., Nakashima, H. (eds.) MMAS 2005. LNCS (LNAI), vol. 3446, pp. 136–148. Springer, Heidelberg (2005)
9. Giunchiglia, F., Mylopoulos, J., Perini, A.: The tropos software development methodology: Processes, models and diagrams. In: AAMAS 2002 Workshop on Agent Oriented Software Engineering (AOSE 2002), pp. 63–74 (2002)
10. Tapia, D.I., De Paz, J.F., Rodríguez, S., Bajo, J., Corchado, J.M.: Multi-Agent System for Security Control on Industrial Environments. International Transactions on System Science and Applications Journal 4(3), 222–226 (2008)
11. Duda, R.O., Hart, P.: Pattern classification and Scene Analysis. John Wisley & Sons, New York (1973)
12. John, C.: Platt Fast training of support vector machines using sequential minimal optimization. In: Advances in Kernel Methods, pp. 185–208 (1999)

Geo-localization System for People with Cognitive Disabilities

João Ramos, Ricardo Anacleto, Paulo Novais,
Lino Figueiredo, Ana Almeida, and José Neves

Abstract. Technology is present in almost every simple aspect of the people's daily life. As an instance, let us refer to the smartphone. This device is usually equipped with a GPS module which may be used as an orientation system, if it carries the right functionalities. The problem is that these applications may be complex to operate and may not be within the bounds of everybody.

Therefore, the main goal here is to develop an orientation system that may help people with cognitive disabilities in their day-to-day journeys, when the caregivers are absent. On the other hand, to keep paid helpers aware of the current location of the disable people, it will be also considered a localization system. Knowing their current locations, caregivers may engage in others activities without neglecting their prime work, and, at the same time, turning people with cognitive disabilities more independent.

Keywords: Cognitive Disabilities, Mobile Communication, Localization, Orientation, Persons Tracking, Ambient Intelligence.

1 Introduction

Increasing people, elder and sick, is becoming a real problem to the social security systems of developed countries. Indeed, due to the rise of life expectancy and the reduced number of births, the population is getting older [13], therefore requiring more care, which means more costs to the health system's. To reduce the impact of this situation there are a few protection services and facilities, like nursing homes

João Ramos · Paulo Novais · José Neves
Informatics Department, University of Minho, Portugal
e-mail: {jramos,pjon,jneves}@di.uminho.pt

Ricardo Anacleto · Lino Figueiredo · Ana Almeida
GECAD - Knowledge Engineering and Decision Support, ISEP, Portugal
e-mail: {rmao,lbf,amn}@isep.ipp.pt

J.B. Pérez et al. (Eds.): *Trends in Prac. Appl. of Agents & Multiagent Syst.*, AISC 221, pp. 59–66.
DOI: 10.1007/978-3-319-00563-8_8

or caregivers. This type of services involves a significant loss of mobility by the patient. This means that the elder leaves his/her house and/or that a person (that may be a family member) moves in.

To prevent or minimize this independence loss there are smart houses [11], which are considered a good alternative, since they use embedded devices that control the patient health and enable remote access to such data by the physician or other caregiver [3, 12, 5]. But outside his/her home this technology isn't available and the patient can not be monitored.

In the medical arena there have been a number of absorbing developments; indeed some diagnostic techniques have been enhanced and others have been created/discovered. Thereby the well-being of individuals and societies suffered a change, for the better. However, there are some diseases that still do not have a cure, such as the ones known as cognitive disabilities. To this type of people it is very complicated to go out alone, to whom orientation may become a very difficult task and help is needed.

Since 1988 assisted technologies have gain attentiveness. As expected, field experts turned their attention on how it may improve the people's quality of life with cognitive impairments [1]. Nevertheless, a set of devices have been developed, but the problem with such tools is that they were advanced to be embedded in the home environment (*e.g.*, smart houses). Like an ordinary person, people with cognitive disabilities may leave their premises and, once outside, this technology may not able to be used or obtained.

However, there are researchers that have been developing new ways to use the available technology outside a premise, *i.e.*, looking at new tools. Such devices need to be easy-to-use, small, lightweight and resistant, otherwise they may not last for a long time [7, 6].

Our development described in [10] not only enables the positioning for this type of people but also allows the caregiver to know the current location of persons with cognitive disabilities. Thus, the independence of people with cognitive disabilities is increased and the same occurs to their caregiver(s) that may have other work without neglecting the care provision.

In section 2 is presented work developed by other authors. Section 3 describes the proposed system, including all its features. Finally, at section 4 a brief reflection about this work is presented.

2 Related Work

A system that may help people with cognitive disabilities outside their premises was developed by Carmien *et al.* [4]. The authors, based on traditional orientation methods, developed a system that enables human beings to use the public transportation system. The person with cognitive disabilities with a smartphone may make use of the bus to travel from one particular place to another. The reduced number of routes that were drawn in a simple map is surpassed and the user may travel to a greater set of destinations.

This project had two main goals. The former had in mind to provide assistance to the user through just-in-time information about the travel path (including routes and the bus that should be reached for and hold). The extra goal aims to a simple and fast way of a process of communication between the user and a caregiver. The position of each bus is given in real-time by a GPS module previously installed on it. This position is then sent to a server. The route is calculated based on the information presented on the server.

Besides the previous described system the authors also developed a second prototype that did not need a support structure. Using an end-user programming tool the caregiver could create scripts according to the activities to be carried out by the person with cognitive disabilities.

In 2009, Liu *et al.* [9] developed an orientation system for handicapped people. On their study the players freely use the prototype, while being remotely controlled, *i.e.*, the users could walk and explore all the system functionalities.

Liu *et al.* look at two different studies. The former tries to identify which features of the indoor system should be extrapolated to the outdoor one. Combining pictures with overlaid arrows, audio and text messages, the user could successfully attain a particular destination. Outside the premises, the orientation is more complicated since it is a more dynamic environment (*e.g.*, traffic and the determination of the relative position of something or someone passing by), which may turn the system unreliable. The second study examines the usability of landmarks in orientation of people with cognitive disabilities. The obtained results show that there are a few considerations that must be taken into consideration when orientation is provided. For example, a near landmark should be used instead of a marker that is outside the user view (*e.g.*, behind a tall building). This landmark should also be presented as a picture in the user's prototype to help him/her to identify it. If the image is hard to associate to the real landmark, the user may become stressed and confused and the orientation system may fail.

AlzNav from Fraunhover Portugal [8] is an open project that intends to familiarize elder people and people in general within their early stages of dementia. This system presents the positioning to the user through an arrow that rotates like a compass indicating the travel path. AlzNav also provides a localization system that allows the caregiver to send a Short Message Service (SMS) and receive the location of the person with needs of care.

3 System Description

One of the simplest ways to establish communication between caregivers and people with cognitive disabilities is through written messages. As an alternative, cellphones may be used and caregivers act as reminders to people with cognitive disabilities. Nowadays, cellphones are being replaced by smartphones. These devices may have several applications installed that help the user in almost every task of his/her life. However, there is a flaw when referring to applications specifically developed for people with disabilities.

This work describes a system that is being developed for this group of people. It is a localization system that helps not only the patient but also his/her caregiver. The main goal is to provide an application easy-to-use that helps the user that moves from one location to another. The localization capability lets the caregiver be aware of the user's current location. This feature may allow the caregiver to develop another activity without neglecting the care provided to the user.

The framework of this system (Figure 1) is divided into three major parts according to its destinated user: the application for the person with cognitive disabilities (designated by *Cognitive Helper Mobile Solution*, Section 3.1), two applications for his/her caregiver (*Caregiver Applications - Mobile and Web*, Section 3.2) and the server.

The server is composed by two modules: database and *Communication Software*. The database stores all the important data that is necessary for the correct operation of the system (like usernames and locations). The *Communication Software* ensures the communications established between the applications and the server.

3.1 Cognitive Helper Mobile Application

The person with cognitive disabilities has access to a mobile application that runs on Android Operative System, aiming at two main objectives: route the user so he/she may end at the pretended target (without getting lost) and locate the user, so caregivers may be aware of his/her actual location.

The detailed framework of this application is present in Figure 2 and it is divided into three parts. The localization of the user is retrieved through a smartphone's GPS module and by an Inertial Navigation System (INS) presented in [2]. The system, once it get the user location, may start the routing since it is possible to calculate the journey between the present location and the destination point.

To turn easier the user routing it is used augmented reality (Figure 3). To create this specific environment it is necessary to use the image captured from the smartphone's camera and using the sensors of the device (accelerometer and gyroscope); it is therefore possible to figure the direction the user is pointing the device. These three elements enable the augmented reality environment, letting the application know to where the user is turned to. A simple green arrow shows the travel path and the distance to be traveled to the next turn. With the localization and augmented reality is then possible to orientate the user and ensure that he/she is moving correctly.

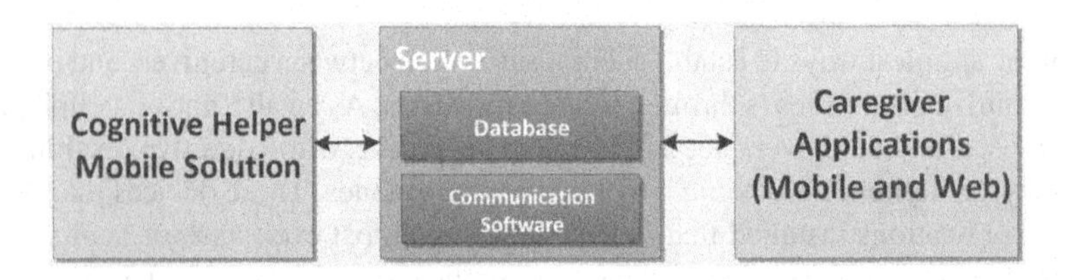

Fig. 1 Simplified framework of the system

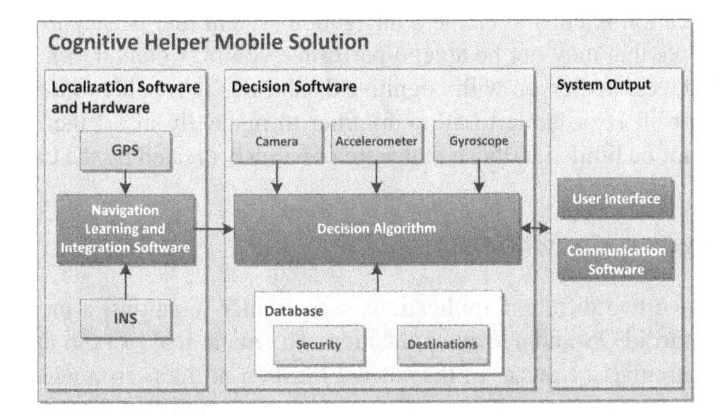

Fig. 2 Detailed framework of the mobile application for people with cognitive disabilities

The *Decision Algorithm* (Figure 2) is responsible for this last task, creating alerts whenever necessary. If the user is at an intersection and is confused on where to go then he/she has to horizontally rotate the smartphone from one side to another and the green flag appears when the smartphone is pointing to the right path.

To interact with the user it is required to use an interface. This module shows, in an interactive way, the information to the user. The selection of a destination is executed through few menus and options. According to the type of destinations stored on the database the user may choose a fixed destination (more common location, like work place or home) or a normal one (generic location, like the mall). To keep this information confidential it is necessary to securely store this information in the database and guarantee that only the user has access to it.

The *Communication Software* establishes the connections between the application and the server. Through this module the application may update the user destinations, update his/her position or receive/send messages from/to his/her caregiver.

Fig. 3 Orientation system using augmented reality

This application intends to create a navigation system that is easy-to-use, especially by people that may not be able to perform a complex mental task. When the caregiver believes the person with cognitive disabilities is capable to correctly use the application, it is possible to allow him/her to manually insert the destination address and not be limited to those that were previously created by the caregiver.

3.2 Caregiver Applications

Caregivers have two different applications with similar functions: a mobile application for Android OS and a Web application. The main goal of both applications is to let the caregiver be aware of the current location of the person with cognitive disabilities; it also presents all the walking paths accomplished from the starting point. The Web application has more capabilities since it is through it that a caregiver registers himself/herself on the platform and creates/adds usernames for the person with cognitive disabilities by whom is responsible.

The framework for these applications (Figure 4) is divided into four modules. The *Communication Software* ensures the transmission of the information between the applications and the server. This data includes not only caregiver's personal information (like his/her username, password, name) but also important data of the patient that he/she is responsible for (like location points, name, destinations stored on the database).

The caregiver is also in charge of the creation of the destination points (through *Travel Path Designer*) that may be used by the patient. This creation may be done by two different methods, *i.e.*, either directly by selecting a point on a map or by searching a location through its address and, if necessary, adjusting the point on the map. Besides the creation of destinations it is the caregiver that specifies if they are starred (which is considered favorite and will be more used) or common ones.

Notifications Creator enables the caregiver to send simple messages of *Yes* or *No* type. This feature allows an easier and fast communication between caregivers and the people with cognitive disabilities. This feature may be useful when the caregiver is watching the traveling path, when it sees, for example, that the patient is passing by a grocery and asks if he/she can buy fruit. Instead of calling and pausing the

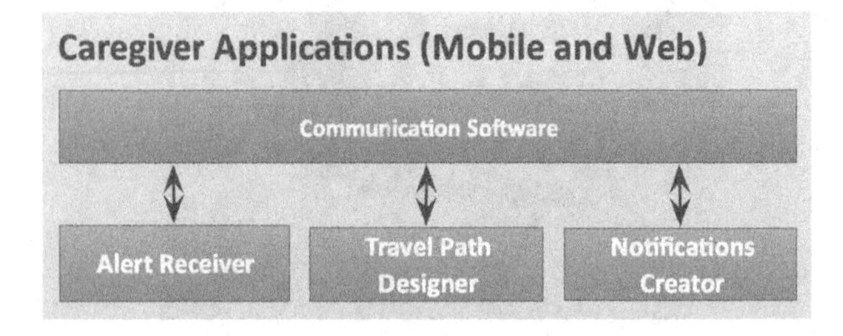

Fig. 4 Detailed framework of the applications for caregiver

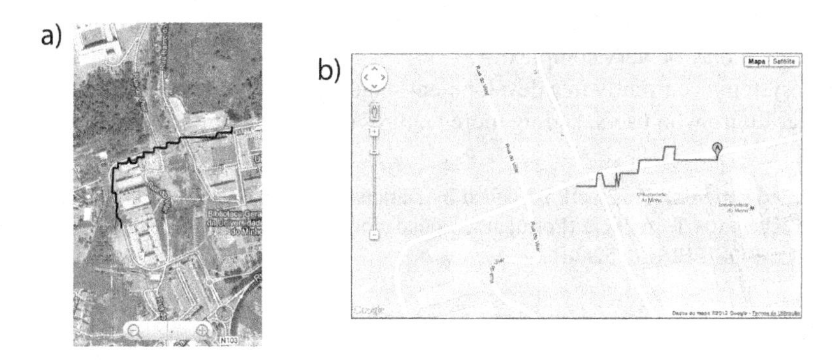

Fig. 5 Location system - a) Android application; b) Web application

navigation, this feature intends to enable a fast answer without interfering with the person routing.

The last module, *Alert Receiver*, receives all the alerts generated by *Cognitive Helper Mobile Solution*. This module informs the caregiver about any events created by the person with the cognitive disabilities, *e.g.*, if the patient successfully reached his/her destination.

Figure 5a) represents the mobile application and shows a travel path that was already done by the person with cognitive disabilities. Figure 5b) represents the Web application and shows a travel path that is being done (the line is updated when the person with cognitive disabilities changes position). To indicate starting and ending points there is a marker on each one.

4 Conclusion and Future Work

Medicine allied to technology is being improved in the last years, but there are some diseases that still do not have a cure. Cognitive disabilities are an example of this situation and physicians tries to reduce the disease progression through medicines.

When it is diagnosed, the patient usually suffers from a reduction of his/her independence. This person is then not allowed to go outside his/her home alone and having a normal live. One of the major problems to people with cognitive disabilities is the lack of orientation, so the risk of getting lost once they go outside alone is very high.

This problem is being studied by several researchers that have been developing some applications/prototypes that route the user when he/she is outdoors.

The lack of orientation of these patients is then having a solution but caregivers still have a problem: when they are not present how do they know where the person with cognitive disabilities is? To answer this question the proposed system is being improved. Besides being an orientation system it is also a localization system.

The use of augmented reality in the orientation system turns it easier to use and surpass the limitations of similar systems, since it does not need static pictures that must be in the same landscape as that of the user, otherwise the way to understand the situation may be very complex.

The system is currently in a development stage and it is expected to do field tests in a near future with users, adding more features (*e.g.* detect preferred routes).

Acknowledgements. This work is funded by National Funds through the FCT - Fundação para a Ciência e a Tecnologia (Portuguese Foundation for Science and Technology) within project PEst-OE/EEI/UI0752/2011.

References

1. Alper, S., Raharinirina, S.: Assisitive Technology for Individuals with Disabilities: A Review and Synthesis of the Literature. Journal of Special Education Technology 21(2), 47–64 (2006)
2. Anacleto, R., Figueiredo, L., Novais, P., Almeida, A.: Providing location everywhere. In: Antunes, L., Pinto, H.S. (eds.) EPIA 2011. LNCS, vol. 7026, pp. 15–28. Springer, Heidelberg (2011)
3. Augusto, J., Mccullagh, P.: Ambient Intelligence: Concepts and applications. Computer Science and Information Systems 4(1), 1–27 (2007)
4. Carmien, S., Dawe, M., Fischer, G., Gorman, A., Kintsch, A., Sullivan, J.F.: Socio-technical environments supporting people with cognitive disabilities using public transportation. ACM Transactions on Computer-Human Interaction 12(2), 233–262 (2005)
5. Carneiro, D., Novais, P., Costa, R., Gomes, P., Neves, J.: EMon: Embodied Monitorization. In: Tscheligi, M., de Ruyter, B., Markopoulus, P., Wichert, R., Mirlacher, T., Meschterjakov, A., Reitberger, W. (eds.) AmI 2009. LNCS, vol. 5859, pp. 133–142. Springer, Heidelberg (2009)
6. Dawe, M.: Desperately seeking simplicity: how young adults with cognitive disabilities and their families adopt assistive technologies. In: Proceedings of the SIGCHI Conference on Human Factors in Computing Systems, CHI 2006, pp. 1143–1152. ACM (2006)
7. Dawe, M.: "Let Me Show You What I Want": Engaging Individuals with Cognitive Disabilities and their Families in Design. Technology, pp. 2177–2182 (2007)
8. Fraunnhover Portugal: AlzNav (2012), `http://www.fraunhofer.pt/en/fraunhofer_aicos/projects/internal_research/alznav.html`
9. Liu, A.L., Hile, H., Borriello, G., Kautz, H., Brown, P.A., Harniss, M., Johnson, K.: Informing the Design of an Automated Wayfinding System for Individuals with Cognitive Impairments. Cognition 9, 1–8 (2009)
10. Ramos, J., Anacleto, R., Costa, Â., Novais, P., Figueiredo, L., Almeida, A.: Orientation System for People with Cognitive Disabilities. In: Novais, P., Hallenborg, K., Tapia, D.I., Rodríguez, J.M.C. (eds.) Ambient Intelligence - Software and Applications. AISC, vol. 153, pp. 43–50. Springer, Heidelberg (2012)
11. Sadri, F.: Multi-Agent Ambient Intelligence for Elderly Care and Assistance. In: Aip Conference Proceedings, vol. 2007, pp. 117–120. Aip (2007)
12. Stefanov, D.H., Bien, Z., Bang, W.C.: The smart house for older persons and persons with physical disabilities: structure, technology arrangements, and perspectives. IEEE Transactions on Neural and Rehabilitation Systems Engineering 12(2), 228–250 (2004)
13. United Nations, Department of Economic and Social Affairs, Population Division: World Population Prospects: The 2010 Revision, New York (2011)

Adding Sense to Patent Ontologies: A Representation of Concepts and Reasoning

Maria Bermudez-Edo, Manuel Noguera, Nuria Hurtado-Torres,
María Visitación Hurtado, and José Luis Garrido

Abstract. At present, information regarding patents is usually represented and stored in large databases. Information from these databases is commonly retrieved in the form of files with a CSV- or XML-based codification but with little semantics that enable the inference of further relationships among patents. In these databases, each patent is associated with a technological field by a code for a hierarchical classification. Although the codes assume a hierarchical classification approach, inclusion/subsumption relationships are not explicitly specified such that computers can process them automatically. This paper presents an approach to automatically translate the hierarchies found in the patent classification codes into ontologies of concept hi-erarchies. This proposal also enables the automatic inference of implicit knowledge of patent information. A case study is presented to illustrate the applicability of the proposal.

Keywords: Ontology; OWL; Patent; International Patent Classification.

1 Introduction

Patents are legal documents that protect the rights of the inventor of an industrial property. A patent document provides valuable information related to the patent innovation, such as, the firm that generates it, location, date, technological field and information about the other patents that it cites. The technological field is

Maria Bermudez-Edo · Manuel Noguera · María Visitación Hurtado · José Luis Garrido
University of Granada. Department of Software Engineering,
E.T.S.I.I., c/ Saucedo Aranda s/n, 18071 Granada, Spain
e-mail: {mbe,mnoguera,mhurtado,jgarrido}@ugr.es

Nuria Hurtado-Torres
University of Granada. Department of Management,
Business School, Campus Cartuja s/n, 18071 Granada, Spain
e-mail: nhurtado@ugr.es

J.B. Pérez et al. (Eds.): *Trends in Prac. Appl. of Agents & Multiagent Syst.*, AISC 221, pp. 67–75.
DOI: 10.1007/978-3-319-00563-8_9

widely used in searches of the databases to determine the field(s) in which a firm may infringe upon another company's industrial rights or where there is an existing gap in the technology in which a company could innovate. All of these data that describe or are related to patent documents are called patent metadata [1].

Patent documents are usually stored in large databases that exhibit a rigid structure. Likewise, these databases often use different data structures that make it difficult to automatically and efficiently process the information contained therein.

Several works have proposed the use of an ontology-based approach to represent patent metadata using the Web Ontology Language (OWL) (e.g., [2, 1]). The main objective of these works is to provide a semantically well-defined and homogeneous representation for the major types of patent metadata. The use of ontologies enables the representation of knowledge and allows for the identification of context and dependency information more easily than using database-centric structures and interfaces [1]. Previous patent ontologies have also included technological field codes in their concept ontologies; however, they do not fully exploit the formal representation of the patent code classification hierarchies of these technological fields.

This paper proposes to enrich the previously proposed ontologies by providing additional meaning to the patent classification codes of technological fields, by representing the hierarchy of the technological codes. This additional meaning enables to analyze the technological activities to identify new context and dependency information by means of description-logics-based reasoning. However, in the patent databases, the technological codes hierarchy is not explicitly described, and it cannot be used in automatic processing by computers. Furthermore, when a new patent classification code is introduced, the database must be changed as well as the applications running on top of the databases. This paper proposed a more flexible and easier way to evolve and maintain patent information repositories so that adding a new code only involves changing the patent ontology. This approach allows the automatic creation of ontologies from the patent databases. To illustrate the applicability of our proposal, this paper presents a practical application in which new information is inferred from the hierarchy of concepts. A case study demonstrates an automatic mechanism for reclassification of patents when a new patent classification code appears.

The remainder of the paper is organized as follows. Section 3 describes the related work. Section 2 explains the translation of hierarchical patent codes into ontological hierarchies of concepts. Section 3 presents the case study. Finally, Section 4 concludes the paper by discussing the contributions of the research.

2 Related Work

The most prominent examples of patent metadata ontology are Patexpert [1] and PatentOntology from Stanford University [2]. Patexpert was created to bring patents from all patent databases into a common format and to provide them with semantic meaning. PatentOntology merges information from patents with information from patent court cases. Both ontologies make use of the a technological classification from patents, but have not implement a full hierarchy of concepts implicit in the technological codes. This field is widely used to delimit the scope of the searches and is one of the most used items of patent metadata [3]. These ontologies basically define a concept for each patent code without leveraging further reasoning capabilities. Specifically, these proposals do not take into account the context of the intermediate parts of each technological code.

3 Analysis of Hierarchical Classification Codes

In this section, we present an analysis based on the hierarchical patent classification codes ontologies. The method begins by splitting the codes for the technological fields into structural parts (pieces of codes) that aid in inferring a hierarchy of concepts as a final output. To automatically create a classification hierarchy of concepts in the patent ontology, this approach includes the following steps:

1. Study the International Patent Classification (IPC) codes and identify certain restrictions.
2. Provide a mechanism to address these features for using a patent ontology for the analysis of the hierarchical classification code.
3. Automatically populate the patent ontology.

3.1 Restrictions of the Patent Classification Codes

Importance of the Parts of the Codes

The first relevant feature found in the IPC codes is related to the meaning of parts of the codes. When dividing the code into its parts, the meaning of the individual parts could be different although their representations are the same. For example, the IPC code *H03K3/03* is defined as "*dealing with astable circuits for generating electric pulses*". This code can be split into the following parts: section (*H*), class (*03*), subclass (*K*), main group (*3*), subgroup (*03*). In this example, the class and the subgroup have the same representation (*03*), but they have different meanings because of their position in the code. One represents the class ("*basic electronic circuits*"), and the other represents the subgroup ("*astable circuits for generating electric pulses*"). Therefore, when translating IPC codes into a hierarchy of concepts, one restriction is that the same individual (in the example, *03*) has different meaning depending on the part of the code to which it belongs (in the example, the class or the subgroup).

Importance of the Context of the Codes

Another relevant feature found in the study of IPC codes is that the meaning of the representation of one part of the code depends on the previous parts of the code. For example, the IPC code *H02K3/02* and IPC code *H03K3/03*, both of them have the subclass (*K*), although the meaning of this subclass is different in each case because the meaning of the subclass depends on the previous parts of the code. This fact must be also taken into account when translating IPC codes into a hierarchy of concepts; that is, the need to put the code into context to understand the meaning of each part. In particular, it is necessary to consider not only the value of a particular part of the code, but also the values of the previous parts.

3.2 HTCOntology: A Proposed Transformation of Hierarchical Codes into Concepts

This subsection provides a method for translation of the hierarchy found in the patent classification codes into a hierarchy of concepts of an ontology.

First, we have designed and implemented a hierarchical classification in the ontology with all of the IPC sections (IPC_A, IPC_B, IPC_C, IPC_D, IPC_E, IPC_F, IPC_G and IPC_H). Next, for each IPC section, we have created subclasses for all of the possible IPC classes of the corresponding section (e.g., IPC_H03). In the same way, we have created the IPC subclasses (e.g., IPC_H03K) and main groups (e.g., IPC_H03K03) (see Figure 1) and the subgroups should be the individuals (e.g., H03K03_03), creating a large hierarchy of classes (HCOntology). The ontologies in the example have been created with the ontology editor Protégé [4]. Thus, we make the hierarchy found in the patent classification codes explicit and enable automatic processing, filling the gaps in the previous patent metadata ontologies.

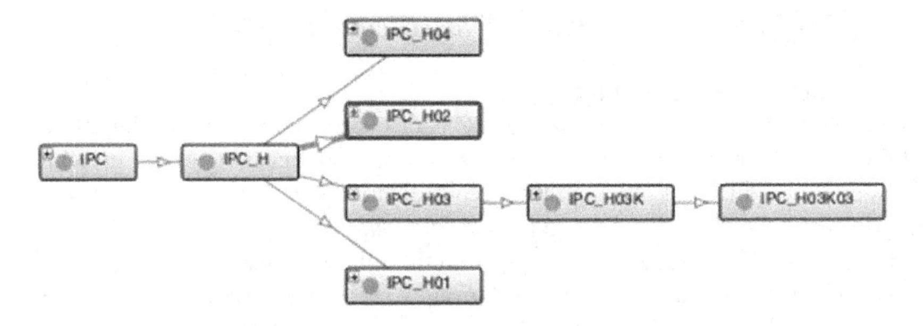

Fig. 1 HCOntology in Protégé for the IPC code H03K03

This design is intended to address the two relevant features mentioned above in this section. The hierarchy of classes in HCOntology solves the first restriction (Subsection 2.1) because one part of the code is a child class belonging to the previous part of the code, and therefore, the meaning of both parts are different.

The proposed HCOntology also addresses the second restriction. In the example of Subsection 2.1, the class IPC_H03K inherits the meaning of the superclass IPC_H03, and another class, IPC_H02K, inherits the meaning of the superclass IPC_H02. Therefore, the meanings of the two codes are well differentiated because the hierarchy of classes takes into account the values of the previous parts of the code.

Furthermore, HCOntology allows reasoning and exploiting the semantics of its hierarchy to create relationships between patents with the same parent technological field. For example, the individual H03K03_03 inherits all of the properties from its parent classes, IPC_H03K03, IPC_H03K, IPC_H03 and IPC_H. Hence, this individual will be found in the searches of IPC codes H03 because it belongs to this parent class.

3.3 Populating the Patent Ontology and HTCOntology

Patent databases available on-line, provide their results in different formats, one of which is XML. We have implemented an automatic mechanism that populates the ontology from the query responses of the patent databases. Following [5], with a style sheet, we created the correspondence between the XML tags and the OWL classes, and therefore extracted the individuals used to populate the ontology. Additionally, this automatic mechanism allows the extraction of each part of the code one-by-one, navigates inside HCOntology, and inserts the IPC code in the class to which it belongs. Figure 2 provides an overview of the translation scheme [6]. The style sheet developed in this work allows population of the ontology automatically. Even when new codes appear, the style sheet can create the new classes and individuals automatically, without any further modification.

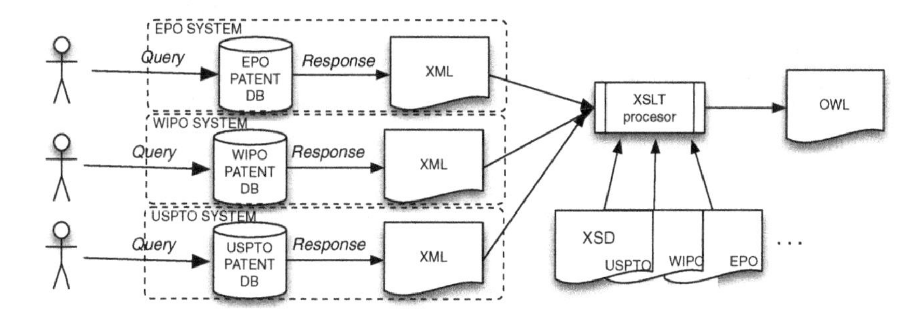

Fig. 2 Method Overview

4 Case Study: Introducing a New Patent Code

The classification hierarchy of concepts can be exploited to infer new knowledge through a reasoner. As the technology evolves, new fields are discovered and patents must represent these new technological fields with their codes. Usually, the need for new codes appears after a group of patents in a field have been filed and classified in other related technological fields. For example, beginning with the Y02 term "technologies or applications for mitigation or adaptation against climate change", the European Patent Office (EPO) has new ICO (In-Computer-Only) [7] classification codes to cover green patents. The EPO have implemented these ICO codes reclassifying patents by searching for particular IPC codes combined with keywords found in patent documents. In this section, we will present a case study that will show how to automatically reclassify patents when a new classification code appears. Specifically, the case study proposes to reclassify patents into this new technological code (green patents) that have not yet been classified as green patents by the patent authorities. With this approach, we will infer new information (ICO codes) derived from other information available in the patent documents (IPC codes and keywords).

To this end, we use OWL [8] as the ontology language together with the reasoner Pellet [9], and we make use of the Protégé tool for the implementation. We have applied this approach in PatentOntology. PatentOntology does not implement ICO codes and certain terms for the concepts and relationships are misleading. For example, there is a property called "*hasIPCClass*" that links the patent with its IPC codes (individuals) and not the IPC Class (classes) as one might think. Therefore, we create the class ICO and for the sake of clarity, we introduce an equivalent object property called "*hasIPCCode*".

In this case study the reasoner will infer a new re-classification of green patents based on the IPC code H04 (and all the subcodes of H04) and keywords related to "bit reduction" because the energy consumed in the transmission of data in computer networks depends on the number of bits transmitted, among other things, and energy consumption is related to environmental issues. We will link these patents to the ICO code Y02_H04_BitReduction. The ontological representation of hierarchical classification of the IPC codes can serve to reason and infer patents pertaining to Y02 codes. In this case, by only selecting the class H04, all of the patents with IPC codes that begin with H04 will be retrieved. This process involves the following steps:

1. Create the classes: Y02_H04BitReduction (a subclass of Y02_code-pending) and PatentH04BitReduction (a subclass of Patent).
2. Populate the ontology with the individuals with IPC codes beginning with H04 and keyword '*bit reduction*'. Add to them the property *ObjectHasValue* (*hasKeyword bit-reduction*).
3. Define the equivalent classes for the class PatentH04BitReduction (see Fig. 3). With the first equivalent class, as it is shown in description logics [10] in axiom 1, the reasoners will search for the patents with the IPC codes H04 and keyword "*bit-reduction*". The second equivalent class, axiom 2, will add the value Y02_ H04BitReduction to the patents found.

$$PatentY02\_H04\_BitReduction \equiv PatentDocument \sqcap \exists hasICOCode.\{Y02\_H04BitReductionIndividual\} \quad \text{(Axiom 1)}$$

$$PatentY02\_H04\_BitReduction \equiv PatentDocument \sqcap \exists hasIPCCode.H04 \sqcap \exists hasICOCode.\{bit-reduction\} \quad \text{(Axiom 2)}$$

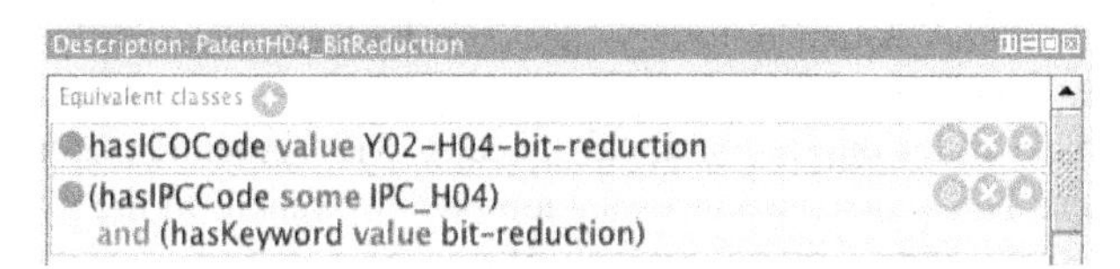

Fig. 3 Equivalent classes that classify patents H04 related to bit reduction as green patents

The reasoner has classified all of the patents with the IPC code H04 and the value "*bit-reduction*" in the class Keyword into this new class of PatentH04_ BitReduction. Hence, when someone searches for Y02 patents, because this individual is in a subclass of Y02, the re-classified patents will show-up. Next, all of the patents that are in the class PatentH04_BitReduction will be linked with the individual Y02_H04-bit-reduction of the class ICO.

We have populated the ontology with more patents, with a number of them meeting the criteria of the case study, such as the patent US20080107132 (see Fig. 4) with the title "Method and apparatus for transmitting overhead information", and classified them with the IPC code *H04J3/24*. The patent addresses a method that attempts to reduce the overhead in the transmission. This IPC code, *H04J3/24*, is defined as "*multiplex communications in which the allocation is indicated by an address*", and therefore an a priori non-green concept is appreciated. However, the patent belongs to the more general code *H04* and has certain keywords such as "*bit-reduction*". Therefore, the patent meets the criteria shown in Figure 3.

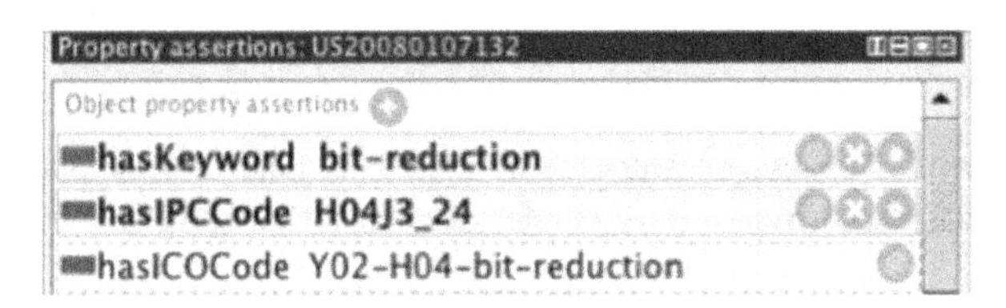

Fig. 4 Patent US20080107132 reclassified

We have run the reasoner Pellet, and the patent has inferred the class Y02-H04-bit-reduction (see Figure 4). Therefore, the reasoner has reclassified patents as green patents making use of part of the classification codes, i.e., using the semantic meaning of the hierarchy found in HCOntology combined with transversal relationships such as keywords and patent classification. Thus, we enable the automatic reclassification of patents based on a combination of patent classification codes and text found in patents (keywords).

5 Conclusions

A patent document provides valuable information about the innovation it protects such as its location, date or technological field. The technological field is one of the most used patent metadata item for innovation research.

While the previous patent ontology literature has mainly used the same technological field codes as in the databases, the analysis of the transformation of these hierarchical codes into hierarchical concepts with semantic meaning remains to be addressed. This work introduces an approach to transform the representation of the IPC patent codes into a hierarchy of concepts. It provides a method that automatically makes that translation, also enabling the automatic translation of future emerging codes without the necessity for any reimplementation. The hierarchy of concepts allows exploitation of the information represented in each part of the technological codes. Particularly, we have exploited the patent ontology, and especially the information represented in the hierarchical classification proposed (HCOntology), to create transversal relationships between concepts automatically processed by computers and without the need for changes in any other application that makes use of the ontology. This proposal enables to automatically update the patent classification of the technological fields (IPC) in an efficient way.

In particular, we have illustrated our proposal with a case study that simulated the introduction of a new code ("green code") into a technological patent classification. This opportunity is especially appealing in the context of steadily increasing technological change, in which multiple technologies (e.g., green technology) are developing quickly. Although we have applied this approach to PatentOntology, it could be used in any other patent ontology. Future studies might explore additional ways to generate and exploit hierarchies of concepts to analyze data in different contexts. Extensions of the proposed patent ontology to analyze other information (not provided in the patent database) are also planned as future work.

Acknowledgments. This research work has been partially funded by the Spanish Ministry of Economy and Competitiveness (project reference number TIN2012-38600).

References

1. Giereth, M., St, A., Rotard, M., Ertl, T.: Application of Semantic Technologies for Representing Patent Metadata. In: 1st International Workshop on Applications of Semantic Technologies, AST 2006, vol. 94, pp. 297–305 (2006)
2. Taduri, S., Lau, G.T., Law, K.H., Yu, H., Kesan, J.P.: An Ontology to Integrate Multiple Information Domains in the Patent System. In: 2011 IEEE International Symposium on Technology and Society, ISTAS, pp. 23–25 (2011)
3. Foglia, P.: Patentability search strategies and the reformed IPC: A patent office perspective. World Patent Information 29, 33–53 (2007)
4. The Protégé Ontology Editor and Knowledge Acquisition System, `http://protege.stanford.edu/`
5. Wanner, L., Brügmann, S., Diallo, B., Giereth, M., Kompatsiaris, Y., Pianta, E., Rao, G., Schoester, P., Zervaki, V.: PATExpert: Semantic Processing of Patent Documentation. Knowledge Creation Diffusion Utilization (2009)
6. Bermudez-Edo, M., Noguera, M., Garrido, J.L., Hurtado, M.V.: Semantic Patent Information Retrieval and Management with OWL. WorldCIST, Portugal (2013)

7. Goebel, M.W.E., Hintermaier, F.S.: Searcher's little helper – ICO index terms. World Patent Information 33, 260–268 (2011)
8. Smith, M., Welty, C., McGuinness, D.: OWL Web Ontology Language Guide (2004)
9. Sirin, E., Parsia, B., Grau, B., Kalyanpur, A., Katz, Y.: Pellet: A practical OWL-DL reasoner. Web Semantics Science Services and Agents on the World Wide Web 5, 51–53 (2007)
10. Baader, F.: The Description Logic Handbook: Theory, Implementation and Applications. Cambridge University Press (2003)

Representation of Clinical Practice Guideline Components in OWL

Tiago Oliveira, Paulo Novais, and José Neves

Abstract. The objective of clinical decision support systems is to improve the quality of care and, if possible, help to reduce the occurrence of clinical malpractice cases such as medical errors and defensive medicine. To do so they need a machine-readable support to integrate the recommendations of Clinical Practice Guidelines. CompGuide is a Computer-Interpretable Guideline model developed in Ontology Web Language that offers support for administrative information concerning a guideline, workflow procedures, and the definition of clinical and temporal constraints. When compared to other models of the same type, besides having a comprehensive task network model, it introduces new temporal representations and the possibility of reusing pre-existing knowledge and integrating it in a guideline.

Keywords: Clinical Practice Guidelines, Ontology, OWL, Clinical tasks, Decision Support.

1 Introduction

Among the healthcare community the occurrence of medical errors and defensive medicine are top concerns [1][2]. Medical errors refer to mistakes during the clinical process that may lead to adverse events, i.e., alterations in a patient's health condition for the worse [1]. They include errors of execution, treatment and planning, and their incidence rates, although not very high, are synonymous with increased spending and loss of life quality for both physicians and patients [2]. To prevent these situations from happening, healthcare professionals often adopt another type of harmful behavior, defensive medicine. Defensive medicine consists in avoiding the treatment of difficult clinical cases to prevent possible

Tiago Oliveira · Paulo Novais · José Neves
CCTC/DI, University of Minho,
Braga, Portugal
e-mail: {toliveira,pjon,jneves}@di.uminho.pt

J.B. Pérez et al. (Eds.): *Trends in Prac. Appl. of Agents & Multiagent Syst.*, AISC 221, pp. 77–85.
DOI: 10.1007/978-3-319-00563-8_10

lawsuits or ordering additional complementary means of diagnostic motivated by the sense of self-preservation of healthcare professionals. This behaviour is also motivated by the overreliance on technological means for diagnostic purposes which in turn is responsible for driving up healthcare costs [3]. If one wants to reduce the impact of medical malpractice, it is necessary to standardize healthcare delivery and provide adequate evidence-based recommendations for clinical encounters [4]. Clinical Practice Guidelines (CPGs) are the current medium of choice to disseminate evidence-based medicine.

According to the definition of the Institute of Medicine (IOM) of the United States (US), CPGs are systematically developed statements that contain recommendations for healthcare professionals and patients about appropriate medical procedures in specific clinical circumstances [5]. They are regarded by healthcare professionals as vehicles through which they can integrate the most current evidence into patient management [6]. However, some limitations are detected in the current format of CPGs. They are available as very long documents that are difficult to consult, since only a small part of these documents are actually clinical recommendations. Moreover, there are some issues concerning the ambiguity that these documents may have [7], namely: the misunderstanding of medical terms (semantic ambiguity); conflicting instructions (pragmatic ambiguity); and the incorrect structure of statements (syntactic ambiguity). Additionally, some forms of vagueness may occur in the text, mainly due to the use of temporal terms (e.g. always, sometimes), probabilistic terms (e.g. probable, unlikely) and quantitative terms (e.g., many, few). A structured format for CPGs that is, at the same time, machine-readable would help to solve these issues by providing an adequate support for guideline dissemination and deployment, at the point and moment of healthcare delivery [8].

This work presents a representation model for CPGs developed in Ontology Web Language (OWL) capable of accommodating guidelines from any category (diagnosis, evaluation, management and treatment) and medical specialty (e.g., family practice, pediatrics, cardiology). As for the organization of this article, it presents in section two the fundamentals about OWL along with some observations concerning the advantages of choosing this knowledge representation formalism over traditional ones, like relational databases. The model, which was named CompGuide, is presented in section three with the different requirements that were taken into consideration during the development phase. Section four presents a discussion about the advantages of the model in comparison with the existing ones and provides some conclusion remarks as well as future directions for this research.

2 Advantages of Ontology Web Language

The OWL Web Ontology Language is a standard developed by the World Wide Web Consortium (W3C) and its current version is OWL 2 [9]. OWL 2 is an update to OWL with increased expressive power with regards to properties, extended support for data types and database style keys. OWL is designed for use by applications that need to process the content of information rather than just presenting information to humans. This formalism facilitates machine interpretability and is built upon other technologies such as XML, RDF and RDF-schema. The advantages of this knowledge representation formalism over RDF are related with the fact that OWL, despite being based on RDF, adds more vocabulary for describing properties and classes (e.g., disjunction, transitivity, simmetry). OWL is composed of three sublanguages: OWL Lite, OWL DL and OWL Full. The sublanguage used for this work was OWL DL and it is named this way due to its correspondence with description logics. An ontology is used to describe the concepts in a domain as well as the relationships that hold between them. To accomplish this task OWL ontologies define three basic components:

- Classes: sets that contain individuals described using formal (mathematical) descriptions that state precisely the requirements for membership of the class;
- Individuals: objects of the domain and instances of classes; and
- Properties: binary relations on individuals that may be used to link two individuals (object properties) or an individual to a data element (data properties).

The advantages of OWL reside in the manner a system uses the information. Machines do not speak human language and, sometimes, there is content that escapes their grasp. For instance, a human knows that in some cases some words are definitely related, although they are not synonyms. A machine does not recognize these relationships, but semantics are important. The idea of OWL is to provide a machine with a semantic context. So the advantage is the creation of a better management of information and descriptions.

In OWL, semantic data is assembled in a graph database that is unlike the more common relational and hierarchical databases (nodes and tables). The relationships in OWL assume a greater importance and are the carriers of the semantic content of individuals. Moreover, it is possible to describe or restrain class membership using these relations and thus accurately delimit their scope. In relational databases this would be a hard task to perform. In fact, there is several software engines developed to reason about the semantic content of ontologies, called reasoners, which check the integrity of the constraints posed on individuals in order to assert if they belong or not to a certain class. Such reasoners are Pellet, FaCT++ and HermiT which are available as plugins for Protégé, the ontology editor and knowledge acquisition system used in this work. The reasoner used in this work was FaCT++. As the objec-tive is the development of a standard machine-readable representation of CPGs, OWL seems to be most suitable.

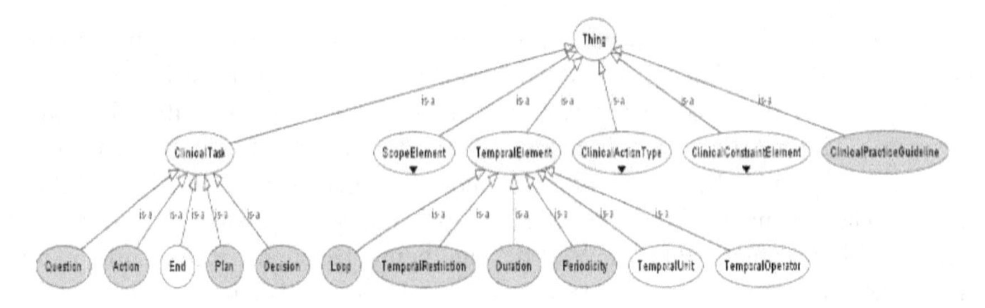

Fig. 1 Diagram of the main primitive classes of the model for Computer-Interpretable Guidelines

3 CompGuide Ontology

There are essentially two ways of developing Computer-Interpretable Guidelines (CIGs): by consulting domain experts in order to get the representation primitives or by researching different CPGs and determine the information needs of clinical recommendations. The method followed in this work was a hybrid one in the sense that it included opinions from healthcare professionals and the observation of guidelines collected from the National Guideline Clearinghouse (NGC).

The main primitive classes of the model are depicted in Fig. 1 and will be described in detail in the following subsections.

3.1 Representation of Administrative Information

As it may be seen in Fig. 1, a CPG is represented as an instance from the *ClinicalPracticeGuideline* class. To keep track of different guideline versions and to provide rigorous descriptions of guideline content and objectives, the individuals of this class have a set of data properties that represent administrative information.

OWL has built-in data types that allow the expression of simple text, numeric values and dates. As such the *string* and *date-time* properties defined for administrative purposes were: *Authorship*, *guidelineName*, *guidelineDescription*, *DateOfCreation*, *DateOfLastUpdate* and *VersionNumber*. There are also additional properties that specify in which conditions and to whom the CPG should be applied, such as *ClinicalSpecialty*, *GuidelineCategory*, *intendedUsers* and *targetPopulation*.

3.2 Construction of Workflow Procedures

CPGs are, essentially, clinical recommendations that are usually presented as sets of tasks that must be performed during clinical encounters, and/or disease management processes. To represent these tasks, CompGuide proposes three main primitive classes, defined under *ClinicalTask*: *Plan*, *Action*, *Decision* and *Question*. The tasks of an individual from *ClinicalPracticeGuideline* are all contained in an individual from *Plan*, to which it is linked through the *hasPlan* object property, as it may be seen in Fig. 2. This is an excerpt of the Standards of Medical Care in Diabetes guideline from the American Diabetes Association, extracted from the NGC. A *Plan* contains any number of instances of the other tasks, including other *Plans*, and it is connected to its first task through the *hasFirstTask* property. In turn, this task is linked to the following task in the workflow by the *nextTask* property and so on. This assures the definition of a sequence of tasks in a manner similar to a linked list, like it is shown in Fig. 2.

The remaining task classes represent different types of activities. Starting with the *Action* class, it represents a step performed by a healthcare agent that includes clinical procedures, clinical exams, medication recommendations and non-medication recommendations. The *hasClinicalActionType* object property connects an *Action* to the different action types defined in *ClinicalActionType*, with appropriate data properties to describe each one.

To express decision moments in the workflow, there is a *Decision* class. The use of this class entails a bifurcation in the clinical workflow and a choice between two or more options. The association of a *Decision* with options and rules is done through object properties that connect them to instances from the *ClinicalConstraintElement* subclasses. The next task in the clinical workflow is selected according to the outcome of the *Decision*. As so, the connection between these tasks is done using the *alternativeTasks* property. This assures that a task is executed instead of another as the result of an inference process guided by trigger conditions.

On the other hand, there may be cases when some tasks must be executed simultaneously, like the procedures of a treatment plan that act synergistically to produce a certain result. For these cases, CompGuide provides the *parallelTasks* property.

The *Question* class is used to obtain information about a patient's health condition, more specifically about the clinical parameters necessary to follow the guideline. In order to fulfill this requirement, there are data properties created to specify the name of the parameter to be obtained and the units in which it should be expressed. Such properties are *string* data types named *Parameter* and *Unit*, respectively.

The *End* class is used to signal the termination of the execution thread that is being followed and to indicate that the guideline reached its finishing point.

3.3 Definition of Temporal Constraints

The importance of time in clinical observations is paramount [10]. When assessing a patient state, a healthcare professional must take into consideration for how long the patient is manifesting his symptoms and try to fit this knowledge in the one he already has about possible causes and solutions. The recommendations of CPGs contain specifications about their temporal execution, namely intended duration, number of repetitions and cycles.

To represent all the temporal constraints, CompGuide provides the *TemporalElement* class. This class contains the two main temporal constructs, *Duration* and *Loop*. The *Duration* class specifies how long a task should last and is defined exclusively for *Plans* and *Actions*. It has a *double* data type property called *DurationValue* where a value for the intended duration of either *Actions* or *Plans* is provided. The *TemporalUnit* class, also defined under *TemporalElement*, contains individuals that represent the different time units in which the *Duration* can be expressed, namely *second*, *minute*, *hour*, *day*, *week*, *month* and *year*. In the *Loop* class it is possible to define cycles for the executions of certain tasks (*Plans* and *Actions*). Each instance of *Loop* has a data property called *RepetitionValue* which is an integer that expresses the number of repetitions that a group of tasks is subjected to. Moreover, each instance also has a *hasPeriodicity* object property that connects it to individuals from the *Periodicity* class (another subclass of *TemporalElement)* which has the appropriate constructs, namely the *hasTemporalUnit* object property and the *PeriodicityValue* data property, to define the regular intervals at which the task is repeated.

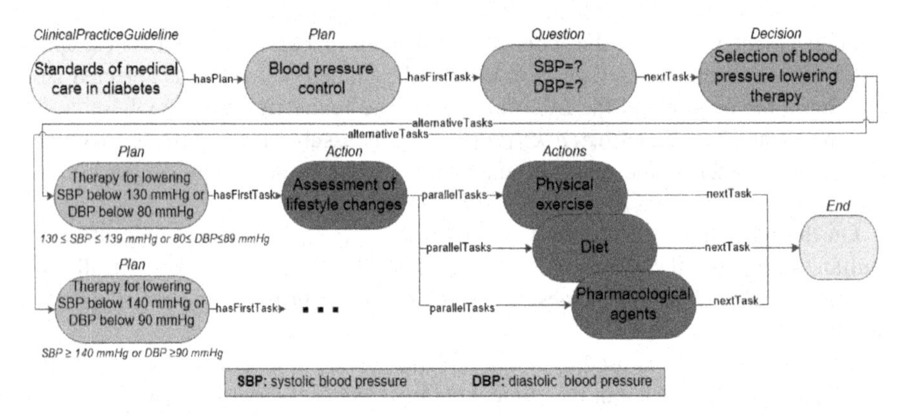

Fig. 2 Excerpt of the Standards of Medical Care in Diabetes clinical guideline from the American Diabetes Association represented according to the CompGuide model

Another feature of the temporal properties is the possibility to define temporal restrictions in clinical constraints. For this purpose one associates a *TemporalRestriction* and a *TemporalOperator* to clinical conditions that must be met for a task to be executed. The temporal operators are based on the theory for quality checking of clinical guidelines by Peter Lucas [11] and include the following individuals:

- *Somewhere in the past*: the condition manifested itself at some point in the past;
- *Always in the past*: the condition manifested itself continuously during a time interval in the past; and
- *Currently*: the condition is manifesting itself during the medical observations.

The *TemporalRestriction* bounds the *TemporalOperator* to a certain period of time. It possesses a *double* data property called *temporalRestrictionValue* and the *hasTemporalUnit* object property. For instance, if an *Action* requires the verification if a patient has been doing is medication correctly for the last 3 months, then *TemporalOperator* is set to *always in the past*, *temporalRestrictionValue* is set to *3* and finally *TemporalUnit* is set to *month*.

3.4 Definition of Clinical Constraints

As it was mentioned previously, a *Decision* implies the choice between two and more options. The association of individuals from *Option* to *Decision* is done by the *hasOption* property. The number of times this property is used in a *Decision* is equal to the number of options the task presents. Each *Option* has a *Parameter* data property and a *NumericalValue* or *QualitativeValue* data properties. The rules that dictate the option selection are provided by the *hasConditionSet* property, linking the individuals from this task to *ConditionSet*. The last one gathers all the necessary conditions through *hasCondition*. In *Condition*, it is possible to define the clinical parameter whose value will be compared, the unit it should be in and the operator that should be used. The *hasComparisonOperator* property connects individuals from *Condition* to *ComparisonOperator*. There, the following individuals were created: *equal to*, *greater than*, *greater or equal than*, *less than*, *less or equal than* and *different from*.

After a medical decision, it is necessary to select the next task in the clinical workflow. Therefore there must be some kind of reference in the tasks that are up for selection (connected by the *alternativeTasks* property) to the possible results of the *Decision*. This is done through the *TriggerCondition* class that also uses the *ConditionSet*. The execution of an activity is triggered when the conditions match the decision output, the selected option. There other classes in *ClinicalConstraintElement* that also use *ConditionSet* in a similar way, namely *PreCondition* and *Outcome*. *PreCondition* is used for all types of tasks to express the requirements of the patient state hat must be met before the execution of a task. For instance, when administering some pharmacological agent it should be known that the patient is not allergic to it. The *Outcome* class puts a restriction to *Plans* and *Actions* that are oriented by therapy goals, like the case of Fig. 2 in which the *Plans* will only be considered completed when the desired levels of SBP and DBP are achieved.

4 Discussion and Conclusions

Although this work draws some inspiration from pre-existing models [8], such as Arden Syntax, PROforma, GLIF3, Asbru or SAGE, it also introduces different views about the definition of clinical constraints, temporal properties, clinical task scheduling and how all these aspects connect with each other. Taking as a reference the oldest, and probably, the most widely (academically) used model, Arden Syntax (now a standard of Health Level 7), which represents knowledge for only one clinical decision, CompGuide provides more expressive power by allowing the definition of a clinical workflow, similarly to GLIF3 and PROforma. However, these models do not have native methods for expressing temporal constraints, using a subset of Asbru temporal language to deal with this issue. Asbru, is, by far, the model that has more temporal constructors and the most complete in this regard, but, at same time, is considered very complex and, in some cases, impractical. The temporal constructs presented in this work are intended to be a compromise between expressivity and complexity that better suits the necessities of clinical decision support systems and thus of healthcare professionals. Another important aspect is the possibility of reusing knowledge from other ontologies in CompGuide by merging the two. This way, the scalability of knowledge is assured.

None of the current formalisms for CPGs is used in a large scale for real context clinical decision support systems. Yet, there is evidence that CIG based decision support could in fact improve the quality of care and address the previously mentioned problems [12]. By using OWL to represent CPGs, one intends to benefit from the advantages of this knowledge representation formalism and if possible, increase the penetration of CIGs into routine medical care.

Acknowledgments. This work is funded by national funds through the FCT – Fundação para a Ciência e a Tecnolo-gia (Portuguese Foundation for Science and Technology) within project PEst-OE/EEI/UI0752/2011". The work of Tiago Oliveira is supported by a doctoral grant by FCT (SFRH/BD/85291/2012).

References

1. Kalra, J.: Medical errors: an introduction to concepts. Clinical Biochemistry 37, 1043–1051 (2004)
2. Chawla, A., Gunderman, R.B.: Defensive medicine: prevalence, implications, and recommendations. Academic Radiology 15, 948–949 (2008)
3. Hermer, L.D., Brody, H.: Defensive medicine, cost containment, and reform. Journal of General Internal Medicine 25, 470–473 (2010)
4. Landrigan, C.P., Parry, G.J., Bones, C.B., Hackbarth, A.D., Goldmann, D.A., Sharek, P.J.: Temporal trends in rates of patient harm resulting from medical care. The New England Journal of Medicine 363, 2124–2134 (2010)
5. Field, M.J., Lohr, K.N. (eds.): Committee on Clinical Practice Guidelines, I. of M. Guidelines for Clinical Practice:From Development to Use. The National Academies Press (1992)
6. Vachhrajani, S., Kulkarni, A.V., Kestle, J.R.W.: Clinical practice guidelines. Journal of Neurosurgery: Pediatrics 3, 249–256 (2009)
7. Codish, S., Shiffman, R.N.: A Model of Ambiguity and Vagueness in Clinical Practice Guideline Recommendations. In: AMIA Annual Symposium Proceedings, pp. 146–150 (2005)

8. Isern, D., Moreno, A.: Computer-based execution of clinical guidelines: a review. International Journal of Medical Informatics 77, 787–808 (2008)
9. W3C: OWL 2 Web Ontology Language Document Overview. W3C (2009)
10. Neves, J.: A logic interpreter to handle time and negation in logic data bases. In: Proceedings of the 1984 Annual Conference of the ACM on the Fifth Generation Challenge, pp. 50–54 (1984)
11. Lucas, P.: Quality checking of medical guidelines through logical abduction. In: Proc. of AI 2003, pp. 309–321. Springer (2003)
12. Novais, P., Salazar, M., Ribeiro, J., Analide, C., Neves, J.: Decision Making and Quality-of-Information. In: Corchado, E., Novais, P., Analide, C., Sedano, J. (eds.) SOCO 2010. AISC, vol. 73, pp. 187–195. Springer, Heidelberg (2010)

Dynamically Maintaining Standards Using Incentives

Ramón Hermoso and Henrique Lopes Cardoso

Abstract. Standards have had much importance in different fields of research in order to assure a certain quality of service in bilateral contracts. More specifically, in multi-agent systems performance standards may be used in order to articulate contracts among partners in environments dealing with uncertainty. However, little effort has been made on how to ensure standards compliance over time. In this work we put forward a learning-based mechanism that attempts to maintain performance standards by applying incentives and/or punishments to agents identified as specialised for certain tasks. We present some empirical results supporting our approach.

Keywords: multiagent systems, standard, incentive.

1 Introduction

A number of research proposals have been made recently concerning the development of infrastructures for supporting interaction in open multi-agent systems. In such systems agents enter and leave the interaction environment, and behave in an autonomous and not necessarily cooperative manner, exhibiting self-interested behaviours. Even when agents establish commitments among them, the dynamic nature of the environment may jeopardize such commitments if agents are not socially concerned enough, valuing more their private goals when evaluating the new circumstances.

Ramón Hermoso
CETINIA, University Rey Juan Carlos - Tulipán s/n,
28933, Madrid, Spain
e-mail: ramon.hermoso@urjc.es

Henrique Lopes Cardoso
LIACC / DEI, Faculdade de Engenharia, Universidade do Porto,
Rua Dr. Roberto Frias, 4200-465 Porto, Portugal
e-mail: hlc@fe.up.pt

J.B. Pérez et al. (Eds.): *Trends in Prac. Appl. of Agents & Multiagent Syst.*, AISC 221, pp. 87–94.
DOI: 10.1007/978-3-319-00563-8_11

Moreover, in open systems one cannot assume that agents will behave consistently along time. This may happen either because of agent's ability or benevolence. In some cases, an agent may not be capable of maintaining a certain behaviour standard throughout its lifetime. In other cases, the agent may intentionally deviate from its previous performance. It is therefore important, when considering open environments, to take into account also the evolution of an agent's internal skills or motivations, besides the dynamics of the interaction environment as a whole.

Looking at the society from a role-specialization perspective, Hermoso *et al.* [3] propose a coordination mechanism, based on role evolution, that assists agents in selecting good partners to whom to delegate specific tasks. The authors look at the agent society and identify "run-time roles" that cluster agents with similar skills for (sets of) tasks. This allows one to identify the role that labels agents most suitable to perform a specific task.

Building on this work, in this paper we associate roles with *performance standards* and address their maintenance: given the dynamics of agents' behaviour, how can performance standards be guaranteed? Two different policies can be used when agents start under-performing. One is to update the role taxonomy and to measure new standards. But assuming that this reorganization may be costly, another option is to influence agents' reasoning by employing incentives, as an attempt to keep them on track.

In Section 2 we put forward a model to establish and adjust incentives in order to maintain standards over time. We present some empirical results in Section 3. Finally, we sum up the paper and point future work in Section 4.

2 Incentive-Based Mechanism to Maintain Standards

While the work in [3] focused on providing a role specialization taxonomy enabling better trust estimations of agents when performing specific tasks, in this paper we assume that such roles may be used to assess performance standards that provide a clearer picture of agents' skills. The path from roles to standards is described in [4]. These measured standards are then to be maintained through an incentive-based policy.

In order to devise an incentive mechanism, we model our interaction scenario according to the well known *principal-agent* model [5, 1] from economics, in which a principal (a service requester) requests an agent (the provider) to perform a specific task. The principal is interested in influencing the efforts that the agent puts when performing the task: efforts correspond to available actions with different execution costs. The exact actions executed by the agent are unobservable to the principal; instead, the latter observes some performance measures. Actions determine stochastically the obtained performance, which is therefore a random variable whose probability distribution depends on the actions taken by the agent. By establishing an incentive schedule, the principal aims at encouraging the agent to choose actions better leading to an intended performance standard.

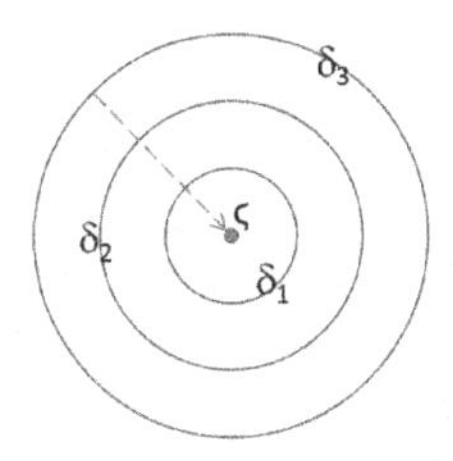

Fig. 1 A standard as a target

2.1 Targeting Standards

Standards are generated using an averaging function applied to task execution outcomes of a group of provider agents (as explained in [4]). Since, according to our model, standards allow requesters to identify expected values for task executions, we consider a standard as a target that agents should meet. Any deviation from the standard is seen as a sub-optimal outcome. Figure 1 illustrates this idea, where ς represents the expected target standard, and each concentric circle labelled with δ_i denotes equidistant performances to the target. These concentric lines highlight the fact that deviations in any direction are considered equally harmful in terms of expected values. The arrow pointing towards the centre discloses the aim of our incentive-based approach: to encourage providers to better target the standard.

We assume each provider has a set of actions at its disposal, each with a cost and a probability function for obtaining different performance outcomes. As follows from Figure 1, an outcome is seen as a *distance* to the standard. This allows us to think of actions as *efforts* the provider puts in when executing a given task: the more effort is invested, the higher the likelihood that the obtained outcome will be closer to the standard. Naturally, expending more effort also means bearing a higher cost.

2.2 Actions, Outcomes and Incentives

More formally, using a finite model for actions and outcomes, we have that:

- The provider has an ordered set of possible actions $\mathcal{A} = \{a_1, ..., a_n\}$, where $a_i \prec a_j$ if $i < j$. This means that $Cost(a_i) < Cost(a_j)$.
- The possible observable outcomes the provider may obtain is an ordered set $\bar{\mathcal{X}} = \{\bar{x}_1, ..., \bar{x}_m\}$, where $\bar{x}_i \prec \bar{x}_j$ if $i < j$ ($\bar{x}_i$ is a worse performance than $\bar{x}_j$). For simplification, we assume that $\bar{x}_i \in [0, 1]$, for all $i \in [1, m]$: each $\bar{x}_i$ denotes the percentage of the target standard that has been achieved.
- There is a probability distribution function for $\bar{\mathcal{X}}$ given an action in $\mathcal{A}$, where $p(\bar{x}_k|a_i)$ is the probability of obtaining outcome $\bar{x}_k \in \bar{\mathcal{X}}$ when performing action $a_i \in \mathcal{A}$. We have that $\sum_{k=1}^{m} p(\bar{x}_k|a_i) = 1$, for all $i \in [1, n]$.

We assume that the monotone likelihood ratio property (MLRP) [1], relating actions with outcomes, holds for every provider. MLRP states that greater efforts are more likely to produce better outcomes: for any $a_i, a_j \in \mathcal{A}$ with $a_i \prec a_j$, the likelihood ratio $p(\bar{x}_k|a_i)/p(\bar{x}_k|a_j)$ is non-increasing in k.

Incentives are specified through an incentive schedule function mapping possible outcomes to values to be collected or paid by the provider: $I : \bar{\mathcal{X}} \to \mathcal{I}$. We take I to be non-decreasing, that is, $I(\bar{x}_1) \leq ... \leq I(\bar{x}_m)$, meaning that higher outcomes must have at least the same incentive as lower ones. Moreover, we look at incentives as producing some change in the utility the agent would get if no incentives were in place; in this sense, $\mathcal{I} = \{\iota : \iota \in [-1, 1]\}$, where positive (negative) values denote percentage increases (decreases) in utility. When $\iota = 0$ there is no incentive in place.

Based on the stochastic model of action outcomes explained above, each provider is taken to be expected utility maximizer. Therefore, when choosing the action to perform it will maximize expected utility [9]:

$$\arg\max_{a \in \mathcal{A}} \mathbb{E}_a = \sum_{i=1}^{m} p(\bar{x}_i|a)u(I(\bar{x}_i)) - Cost(a) \tag{1}$$

where $u(I(\bar{x}_i))$ is the utility the agent gets from obtaining performance outcome $\bar{x}_i$ and consequently incentive $I(\bar{x}_i)$. Function $u : \mathcal{I} \to [0, 1]$ is taken to be strictly increasing. We assume provider agents are risk averse. We define function u using a sigmoid:

$$u(I(\bar{x})) = \frac{1}{1 + e^{-I(\bar{x}) \cdot B + \kappa}} \tag{2}$$

where $\kappa \in \mathbb{R}$ represents a parameter to tune the center of the sigmoid function and $B \in \mathbb{N}^+$ allows us to tune the sensitivity to received incentives.

2.3 Deviations and Responses

Given the previous performance of each provider, on which standards have been defined, we identify two possible causes for agents to deviate from those standards, in the sense that they are not able to meet them anymore. Such causes naturally come to surface from analysing Equation 1: i) action costs have changed, leading an agent to choose actions that stochastically obtain lower outcomes; ii) probabilities for an action's performance outcomes have changed, e.g. due to environmental factors not under the control of the agent, meaning that a specific action is not as effective as before.

These deviations in performance may make a role taxonomy and its previously measured performance standards inaccurate to represent agents' current capabilities. In order to maintain standards, the system may determine and employ an appropriate incentive schedule $I : \bar{\mathcal{X}} \to \mathcal{I}$, which is based on measurable outcomes of task execution. As mentioned before, an outcome is a percentage of the target standard that has been met. Unlike typical approaches in game theory, we do *not* assume any knowledge of the incentive policy maker regarding action costs and probability distributions over outcomes, or provider utility functions. Thus, we see the problem of searching for an optimal incentive schedule as a *reinforcement learning* (RL) [8] problem.

In the following we briefly describe how states, actions and rewards are addressed in the problem faced by the incentive policy maker.

States. The state entails recently obtained performance outcomes. States exhibiting performances farther away from the target standard need to be addressed with stronger incentive policies, while states denoting abidance to agreed standards need no intervention from the policy maker.

Depending on how performance quality is to be interpreted, we may aggregate recent task executions in different ways. In this paper we rely on an average: $perf = \left(\sum_{i=t-\Delta}^{t} \bar{x}^i\right) / \Delta$, where t is the current time step, $\bar{x}^i$ is the outcome obtained at time step i and Δ is the size of the time window, i.e. the number of task executions to consider.

In order to reduce the size of the state space, states are discretized according to the number of levels of deviation that are to be addressed differently, as illustrated in Figure 1. We define a δ parameter specifying in how many intervals to split the distance to target standards:

$$state = \begin{cases} 1 & \text{if } perf = 1 \\ \lfloor perf \cdot \delta \rfloor / (\delta - 1) & \text{if } perf < 1 \end{cases}$$

This function gives us δ different states, represented by values within $[0, 1]$.

Actions. Available learner actions concern incentive schedules I that specify, for any $\bar{x} \in \bar{\mathcal{X}}$, an incentive value $\iota \in \mathcal{I}$. Following [1], each action can be seen as a non-decreasing incentive vector $(\iota_1, ..., \iota_m)$, where m is the number of possible outcomes. In order to reduce the action space, we consider only incentive values in the set $\lfloor \mathcal{I} \cdot 10 \rfloor / 10$ (discrete values with 0.1 steps). Yet, depending on the number of outcomes to consider, this may still give us a huge number of actions to experiment with.

The heuristic we use to tackle with this problem is to explore the action space by generating incentive schedules that consist of minor changes to the currently employed schedule: we step-change one of the incentive values and if needed fix the rest of the schedule to guarantee the non-decreasing property. A *softmax* policy [8] is used to select among the actions considered.

Rewards. An optimal incentive schedule should take into account both the obtained provider outcomes and the cost of applying the incentive schedule. In our approach, these costs are associated with the actual performances that such a schedule has led to, since incentives are paid (if positive) or collected (if negative) according to actual outcomes. Considering that the mechanism does not seek profit, but rather to intervene as least as possible, we sum the absolute values of actually applied incentives when computing the incentive schedule cost.

A reward is computed as a weighted difference between the sum of obtained outcomes and the cost of the incentive schedule. Using weights allows us to define the relative importance of providers' performance and incentive cost.

In RL, $Q(s,a)$ values are computed to determine the expected return for executing action a in state s. We update these values using the simple update rule $Q(s,a) = Q(s,a) + \alpha \cdot (reward - Q(s,a))$, where α is a step-size parameter (we use $\alpha = 0.3$ for the following experimental evaluation).

3 Experiments

We have implemented a simulation environment by using the Repast framework. In order to calculate actual outcomes when a task is requested, providers' behaviour is defined in terms of possible outcomes. In order to do that, we need to set a relationship between efforts and actual outcomes. We have modelled this issue by using beta distributions. There exists a different beta distribution for every different possible effort, in order to be able to calculate actual outcomes. We consider as possible outcomes the set $\bar{x}_1, \bar{x}_2, \ldots, \bar{x}_7$, where $\bar{x}_i$ are different equidistant values in $[0, 1]$. For the sake of simplicity, the number of different efforts available to providers is the same (although it needs not be): $a_1, a_2, \ldots, a_7$. We set the centre value for each effort as the outcome value with the same index: each effort a_i will obtain an outcome modelled as a beta distribution centred in $\bar{x}_i$. In the experiments reported in this paper, all providers share the same beta distributions.

We also need to define a function for effort costs. These costs are used in the provider's decision making (see Equation 1). For that purpose, we use Equation 3 to define different profiles of providers. This means that different providers may have different costs for the same efforts.

$$Cost(a) = \alpha \cdot (\rho + (1 - \rho) \cdot a^{1/\beta}) \tag{3}$$

In this set of experiments we have a heterogeneous population of 100 providers, with random values for α, β and ρ, thus obtaining individuals with a different curve relating efforts to their costs. We set values κ and B to 0 and 10, respectively (see Equation 2). Those values fix the sensibility of providers to incentives in their decision-making processes.

We simulate task requests from customers, one to every different provider in every time step. We show average results from 10 different runs.

In Figure 2(a) we observe how our approach progressively learns an appropriate incentive schedule, which induces providers to behave better: they get progressively closer to the standard. We can also see that this process takes some time, since there exist a high number of possible new (unexplored) incentive schedules that can be generated in each step of the learning process. Once the approach converges to an (almost) optimal achievement of the

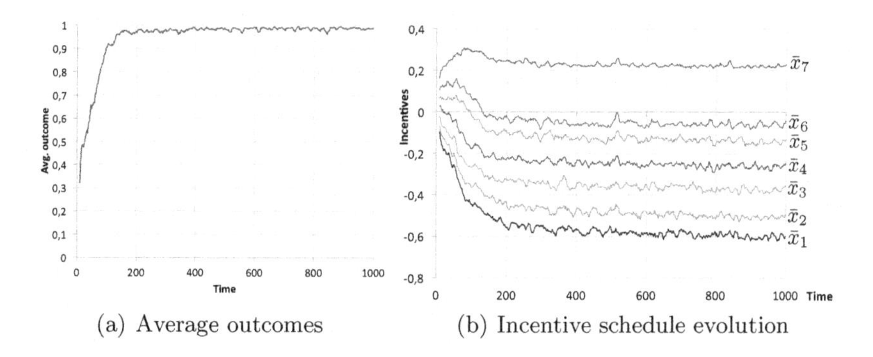

(a) Average outcomes (b) Incentive schedule evolution

Fig. 2 Experimental results

standard, an appropriate incentive schedule makes providers select the most reliable action (in terms of standard achievement).

Figure 2(b) shows the evolution of the incentive schedule employed. Incentives applied to each possible outcome are shown.

4 Conclusions and Future Work

Standards are used as a means to articulate contracts in social interactions. In this paper we have proposed a mechanism that provides incentives to make agents maintain a level of performance as close as possible to the standards. Some possible applications of this approach cover from manufacturing systems, in which agents playing different roles when building a craft are supposed to meet and maintain a standard during their work, to social systems such as ruled electronic markets, where while standards may not be known a priori, they can be discovered at runtime and artificially maintained for the sake of the overall market community.

There are economic approaches also founded on the emergence of standards. Sherstyuk [7] proposes a method to set appropriate performance standards to develop optimal contracts, in which the provider's best choice is to keep the standard through its action. In this paper, however, we are not pursuing optimal performance standards; instead, we are concerned about how to maintain the level of those standards once they have been created.

In the same line Centeno *et al.* [2] present an approach on adaptive sanction learning by exploring and identifying individuals' inherent preferences without explicit disclose of information – the mechanism learns over which attributes of the system should modifications be applied in order to induce agents to avoid undesired actions. In our case, we adhere to a more formal scenario, in which interactions are regulated by means of contracts. Moreover, we assume that the attributes that may be modified by means of incentives are already known by the mechanism.

The approach taken in [6] also assumes that the mechanism knows which attributes it should tweak in order to influence agents' behaviors, namely by adjusting deterrence sanctions applicable to contractual obligations that agents have committed to. The notion of social control employed there is similar to our notion of role standard maintenance; however, instead of a run-time discovered standard, a fixed threshold is used to guide the decisions of the policy maker. Moreover, only sanctions (seen as fines) are used to discourage agents from misbehaving, while here we are also interested in incentivating agents to do their best (by using appropriate actions) while executing the tasks they are assigned to.

We intend to pursue the mechanism presented in this paper, namely by refining the learning model of the incentive policy maker. We also intend to combine the approach with the decision on when to reconfigure the role taxonomy from which standards have been generated.

References

1. Caillaud, B., Hermalin, B.: Hidden action and incentives, Teaching Notes, U.C. Berkeley (2000), `http://faculty.haas.berkeley.edu/hermalin/agencyread.pdf` (accessed)
2. Centeno, R., Billhardt, H., Hermoso, R.: An adaptive sanctioning mechanism for open multi-agent systems regulated by norms. In: Proc. of the 23rd IEEE Int. Conf. on Tools with Artificial Intelligence, pp. 523–530. IEEE Computer Society (2011)
3. Hermoso, R., Billhardt, H., Ossowski, S.: Role evolution in open multi-agent systems as an information source for trust. In: 9th International Conference on Autonomous Agents and Multi-Agent Systems. pp. 217–224. IFAAMAS (2010)
4. Hermoso, R., Lopes Cardoso, H.: Dynamic discovery and maintenance of role-based performance standards. In: Ossowski, Toni, Vouros (eds.) Agreement Technologies. CEUR Workshop Proceedings, vol. 918, pp. 27–41. CEUR-WS.org (2012)
5. Laffont, J., Martimort, D.: The Theory of Incentives: The Principal-Agent Model. Princeton Paperbacks. Princeton University Press (2002), `http://books.google.pt/books?id=Yf1TwtIuNf8C`
6. Lopes Cardoso, H., Oliveira, E.: Social control in a normative framework: An adaptive deterrence approach. Web Intelligence and Agent Systems 9, 363–375 (2011), `http://dx.doi.org/10.3233/WIA-2011-0224`
7. Sherstyuk, K.: Performance standards and incentive pay in agency contracts. Scandinavian Journal of Economics 102(4), 725–736 (2000)
8. Sutton, R.S., Barto, A.G.: Reinforcement Learning: An Introduction. The MIT Press (1998)
9. Von Neumann, J., Morgenstern, O.: Theory of Games and Economic Behavior, 3rd edn. Princeton University Press (1980)

Self-organizing Prediction in Smart Grids through Delegate Multi-Agent Systems

Leo Rutten and Paul Valckenaers

Abstract. This paper discusses a contribution to software and system engineering for smart grids, which comprises multi-agent application domain modeling and delegate multi-agent systems. In this contribution, domain models are software components – agents offering executable services – that become part of the multi-agent software as it will be deployed. These domain models crystalize relevant power engineering knowhow and expertise, thus building bridges between the power engineering and the software engineering communities. The longevity of the real-world counterparts of these domain models ensures their technical feasibility and economic value. By mirroring real-world counterparts throughout their full life cycle, re-configurability is ensured and, in combination with the evaporate-and-refresh mechanisms of the delegate multi-agent systems, even becomes business-as-usual. The paper's main contribution originates from research addressing self-organizing prediction of smart grid operations.

Keywords: Delegate multi-agent systems, smart grid, self-organizing prediction, Holonic systems.

1 Introduction

Smart grid software design and development faces some unique challenges; it is indeed a highly demanding problem domain for software engineering. Conversely, translating and applying the state-of-the-art in software engineering to smart grid development is vital to cope with smart grid complexity, diversity and

Leo Rutten · Paul Valckenaers
IWT department, KHLim, Diepenbeek, Belgium
e-mail: {leo.rutten,paul.valckenaers}@khlim.be

Paul Valckenaers
Mechanical Engineering Department, KU Leuven, Leuven, Belgium
e-mail: Paul.valckenaers@mech.kuleuven.be

J.B. Pérez et al. (Eds.): *Trends in Prac. Appl. of Agents & Multiagent Syst.*, AISC 221, pp. 95–102.
DOI: 10.1007/978-3-319-00563-8_12

heterogeneity. A continued lagging of the state-of-the-art in software engineering (which unfortunately is quite common in industrial automation) would represent a significant loss to society.

A key challenge is bridging the gap between the power engineering and the software engineering communities. Jackson [1] claims that a software development team needs to master the problem domain — in casu, power engineering — to be able to successfully develop mission-critical software systems.

In contrast, Jackson pointed out at the IBM Chair in 1985 in Leuven (B) that the problem domain is the most stable aspect in software developments. Using administrative software as a sample problem domain, Jackson rightfully mentioned that *the employee life cycle* remains largely unchanged during decades, whereas functional requirements (i.e. detailed specifications of the management reports) are likely to change almost on a weekly basis.

This observation equally applies to the smart grid. Grid components will not change overnight. The development and introduction of transformer technology, power generating technologies, power consumption devices, transmission cable materials all require significant amounts of time and effort while the technological changes rarely have an impact concerning properties that impact grid coordination and control. Likewise, the grid itself will not change overnight. Installing or replacing transformers, generators, transmission lines happens slowly from an ICT perspective.

This paper presents a conceptual design on how to capitalize on Jackson's insight and observation by emphasizing problem domain models as an important element in the software systems that are to be developed and deployed for the smart grid. Because of the presence of a multitude of stable elements in the problem domain, crystallizing power engineering knowhow and expertise within software components will be possible, both in the technical and economic sense.

This paper first discusses its scope: the specific problem domain that it addresses within the smart grid context. Next, it present its software system engineering approach and software mechanism (patterns) that enable to adapt to changes in the world of interest. Specifically, a delegate multi-agent system is used to collect, process and disseminate relevant information, which is "not local to a stable element in the problem domain" but needed or relevant for managing the smart grid operations. Finally, the contribution to smart grid research is discussed.

2 Scope

The research discussed in this paper is complementary to smart grid real-time control. It focuses on a time window beyond the reach of grid control systems that manage the grid operations in (hard) real time; *its time window starts from minutes into the future.* On the other hand, its time window is bounded by the build-up of uncertainty when looking farther into the future. Typically, uncertainty about wind and sun renders predictions useless beyond a number of days into the future (i.e. the software system adds nothing to the commonly available statistical information). From a control system perspective, the envisioned software systems aims to bring and keep the grid into a comfortable state when the control performs its task. For instance, intelligent refrigerators refrain from starting their cooling when electricity supply is scarce. The envisaged software system therefore does not compete with control systems (research). Smart grids evolve in dynamic environment and so the control is done relatively to a prediction at short term.

Furthermore, the design addresses the need to anticipate (hours and days into the future) in a smart grid. Indeed, renewable energy sources dictate when they are able to supply power (e.g. wind, solar) or have complex constraints (e.g. CHP) whereas intelligent consumers are able to shift and even adapt their consumption provided that they receive relevant information ahead of time (e.g. to cool down before a peak in demand).

Importantly, the multi-agent design will predict the impact of future interactions. For instance, when e-vehicles intend/plan to start charging their batteries simultaneously, the software predicts the peak in demand, which allows the intelligent consumers to take notice of this prediction (of an undesirable future state), adapt in time and spread their load over a longer time period. This happens in a decentralized, self-organizing manner. Note that the prediction mechanism copes with situations that may never have occurred before.

Moreover, the envisaged system emphasizes precise modeling of whatever is relevant in the problem domain:

- *Full paths*: every entity on the path[1] from energy source to sink is taken into account. E.g., the system will predict the temperature of the transformers in the grid, assuming the current intentions or plans are executed.
- *All relevant attributes of the energy flows*: power, reactive power (cos Φ), balance across phases… will be covered by the prediction mechanism when and where relevant.
- *Control laws and policies*: the prediction mechanism accounts for the prevailing decision and control mechanisms in the grid. In fact, they are treated as elements belonging to the problem domain.
- *Dispatch-ability*: the ability of a consumer or producer to react to external commands or controls. This information enables a control system, for instance, to minimize the impact of a disruption by dispatching the least-effected consumers and producers. It also allows the application to monitor and manage the margins to handle disruptions dynamically.
- *Negligible discretization errors*: the system does not impose time buckets or other discretization that may impact the applicability of its models.

---

[1] This part of the research presently copes with distribution (i.e. the grid is *not meshed at any given instant in time* although grid connectivity may change over time) whereas meshed transmission networks remain to be addressed in future work. Currently, the meshed subnets within a transmission net are considered to be single copper plates.

- *Refresh and evaporate*: the system regular regenerates information, including the predictions, to account for any changes and disturbances.
- *Flexible mechanisms to collect compute and disseminate information*: the design allows application developers to add what they need (e.g. pricing for dispatching rights).

3 Self-organizing Prediction in Smart Grids

This section discusses the system architecture and design that provides a self-organizing prediction service in smart grids while handling grid re-configurations.

3.1 Structural Decomposition

The approach is centered on problem domain modeling, where these models become part of the finally deployed software. Accordingly, structural decomposition comes naturally and, as a consequence, the software system design scales with the size of the underlying grid in the real world. The main classes of problem domain entities are: Resource types; Resource instances; Activity types; Activity instances.

Typical resources are generators like wind mills, consumers like a refrigerator, power transport lines and transformers.

The resource type (model) for e.g. a power transformer mirrors its technical characteristics. An important part of this model reflects how its temperature changes in function of an electric current profile, an initial temperature and the environmental conditions. It also models the impact of this temperature on the life expectancy of this transformer.

The resource instance of such a transformer tracks the state of a physical instance. This model also comprises connections to the resource instances connected to this transformer. Resource instances (models) thus allow to discover the entire grid to which they are connected. Through a regular refresh, any reconfiguration will be detectable. Resource instances also offer a capacity reservation service, which the delegate MAS (multi-agent systems) use to generate the predictions.

Activity type models reflect the manners in which tasks (e.g. ensuring that a refrigerator stays cold enough) can be performed. Note that this typically will be non-deterministic models capable of generating alternative courses of action (e.g. shifting power consumption in time).

Activity instances mirror an actual task, which includes its state but also its intentions. A refrigerating activity instance thus knows the current state (temperature, content) and a prediction of future cooling periods.

Resource and activity instances also model any policies that they apply. Indeed, decision making mechanisms are considered to reside in the domain model. E.g. a dump refrigerator will have a fixed temperature at which it switches cooling on or off. The more intelligent ones will use several delegate multi-agent systems to coordinate on a system-wide scale.

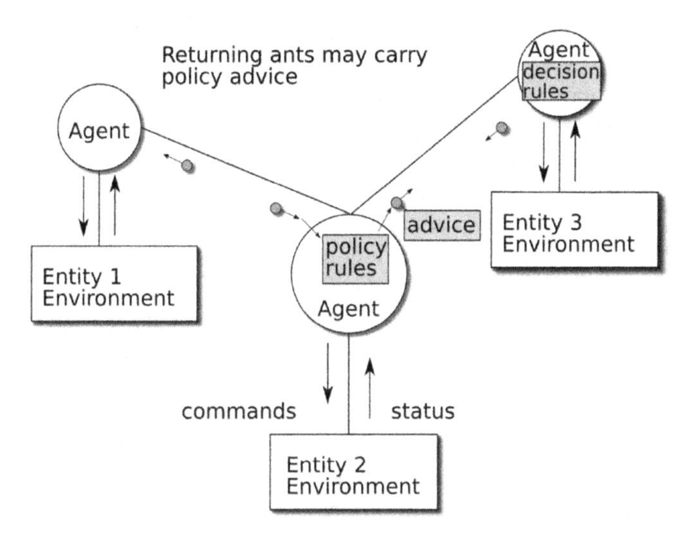

Fig. 1 Agents represent real life entities

The approach makes the distinction between types and instances within the application domain, not the software.

3.2 *Single Source of Truth*

The above structural decomposition applies the *single source of truth principle*, which is closely related to database normalization. Importantly, whenever something changes in the world of interest, only a single model needs to be updated (preferable by an automated tracking mechanism).

This principle simplifies operations radically and eliminates many sources of errors and problems. For instance, any attempt at double bookings will be resolved by the allocation policy of the affected resource. Indeed, its resource instance model is the only source of truth concerning its state and intentions (about future states).

This approach, however, leaves an issue unresolved: how to collect, compute and deliver information that is not collocated with a domain entity (resource or activity)? How does the system predict the future states of resources and trajectories of activities, which obviously involve other resources and (interactions with) other activities? That is addressed through delegate multi-agent systems [2],[3].

3.3 *Delegate MAS*

A D-MAS (delegate multi-agent system) is a software pattern — implementing a mechanism to make non-local information remotely available (i.e. where the information is needed). A forget-and-refresh mechanism ensures that such information is updated regularly such that changes and disturbances are accounted for within a short delay. This pattern is a bio-inspired design — by ant colony food foraging — turning the above single-source-of-truth model of the grid into a major part of the overall solution.

A delegate MAS consists of a stream of lightweight agents called ants or ant agents, which perform an information gathering and/or dissemination task on behalf of their creator (typically a resource or activity). The delegate MAS preserves the computational efficiency of its natural inspiration (i.e. complexity of computations and communication is low-polynomial).

Without being exhaustive, every grid activity instance utilizes an exploration D-MAS and an intention-propagating D-MAS. The exploring D-MAS discovers and collects possible solutions for the task-at-hand. The intention-propagating D-MAS reserves capacity at the selected resources (i.e. full path from generator to consumer). ***These reservations make the expected future interactions (contention) visible***. Forget-and-refresh allows the activities to adapt as the situation evolves and develops. The traveling of ants is shown in fig. 1. This figure also shows how returning ants carry extra advice added by intermediate resources. Consumers use this advice to refine their decision taking rules.

3.4 Smart Grid as a Dynamic Environment

As stated above, to cope with dynamic changes and disturbances in the smart grid, information has only a limited life time. It evaporates (is forgotten) and must consequently refreshed. Resource capacity reservations are regularly reconfirmed by intention-propagating ants. During such refresh, any changes are discovered (e.g. an overly optimistic windmill unable to deliver the promised power). Likewise, the exploration D-MAS may discover superior solutions relative to the current intentions.

As a consequence, activities may need or want to change intentions. If such change happens too easily, the predictions will suffer (as they reflect these intentions). If such change happens too rarely, system performance suffers when opportunities are lost and disturbances are not addressed. Hadeli [4] has investigated this matter and proposed mechanisms to balance responsiveness against stability.

4 Software Engineering for Smart Grids

This paper addresses the challenge of building bridges, in a figurative sense, between the power engineering and the software engineering domains. It comprises the elaboration of executable software models that capture the relevant domain knowhow and expertise to address this challenge in a durable manner. The envisioned software systems do not longer rely exclusively on specialized — in the application domain —human development teams (cf. Jackson in [1]) to account for the specific nature of the smart grid domain. And, the resulting systems are able to account for power grid concerns that are ignored elsewhere, such as by electronic markets matching producer and consumer without accounting for transmission or distribution.

The envisioned solution includes the delegate MAS pattern allowing to use a single-source-of-truth design. Single-source-of-truth permits to extend every entity in the grid with a software extension as its only point of reference whilst the non-local information is generated and refreshed by the appropriate delegate MAS.

For future research in software engineering for the smart grid, this implies the elaboration of executable domain models and delegate MAS variants. This is analogous to a map technology where the map's legend needs to be filled with all elements occurring in the world-of-interest. In fact, the research aims to build a base, mirroring the power grid, on which applications are realized and maintained swiftly and with little effort, analogous to navigation on top of a suitable map. Another analogy is to consider the domain models and delegate MAS as a 'grid operating system' responsible for a predictive resource allocation layer on top of which applications execute.

5 Conclusion

5.1 Contribution

This paper discusses how to build bridges between the power engineering and the software engineering communities. It applies a core achievement in software engineering — domain modeling — and discloses the extent of what can be achieved with this approach in smart grids:

- Single source of truth designs capable of collocating their problem domain models with their real-world counterparts, offering their services inside deployed software systems. Thus, the models may accompany their real-world counterparts where a single model can be reused wherever its domain entity exists.
- A delegate MAS pattern to collect, compute and disseminate relevant information that cannot be collocated with a long-lived domain entity.
- Delegate MAS combined with domain model services that generate, and regularly update, predictions of resource states and activity trajectories accounting for properties and attributes as needed. Among others, power transport and transformation between generator and consumer are accounted for.

Overall, the research aims at and enables to capture power engineering expertise into software, complementing expertise in brains of software developers.

5.2 *Future Research*

The current research results cover only a small subset of the problem domain models needed to cope with and contribute to smart grid operations. The base mechanisms have been established but the collection of executable models is incomplete. Moreover, one important problem domain aspect remains future research: meshed power networks. The base architecture puts forward staff holons whenever a non-local concern needs addressing. The smart grid needs such a holon that computes the available capacities in between connection points of a meshed network. This staff holons needs to recompute and refresh this information regularly.

Finally, the delegate MAS pattern needs to be applied to building up commitments gradually within the time window. Farther in the future, commitments are weak while exploration dominates. Closer to the present, these commitments have to be firm. In the presence of multiple independent organizations, commitments involve pricing and contracts. Pricing needs to induce proper behavior (i.e. prevent cheating and manipulation). For a gradual commitment buildup, pricing mechanisms need to cover options, cancelations, dispatch-ability rights, etc. Here, market deregulation will be key.

References

1. Hinchey, M., Jackson, M., Cousot, P., Cook, B., Bowen, J.P., Margaria, T.: Software engineering and formal methods. Commun. ACM 51(9), 54–59 (2008)
2. Holvoet, T., Weyns, D., Valckenaers, P.: Patterns of delegate mas. In: SASO, pp. 1–9. IEEE Computer Society (2009)
3. Van Dyke Parunak, H., Brueckner, S., Weyns, D., Holvoet, T., Verstraete, P., Valckenaers, P.: E pluribus unum: Polyagent and delegate mas architectures. In: Antunes, L., Paolucci, M., Norling, E. (eds.) MABS 2007. LNCS (LNAI), vol. 5003, pp. 36–51. Springer, Heidelberg (2008)
4. Hadeli, Bio-inspired multi-agent manufacturing control systems with social behaviour, Ph.D. dissertation, KULeuven, Faculty of Engineering (2006)

Holonic Recursiveness with Multi-Agent System Technologies

Sonia Suárez, Paulo Leitao, and Emmanuel Adam

Abstract. Recursive Systems are often needed or recommended for automatically deploy or build a software system composed of multiple entities on a large network, or on distributed locations, without a central control. We use the recursive definitions proposed in holonic systems, as the recursiveness is an elemental structural property of their structure. Holonic systems are used to be implemented using MAS technologies (platforms) on the basis of the shared functional properties of autonomy and cooperation of agents and holons. This paper discusses the adequacy of the actual MAS technologies for holonic recursiveness implementation. A comparison of MAS platforms is done through a framework that will guide the decisions of designers and developers.

1 Introduction

The development of a solution for supporting large and complex systems requires to follow a non-centralized and recursive approach. Herbert Simon (see "Sciences of the Artificial", [11]) proposed to use a recursive and hierarchical model of problem decomposition, and to follow a bottom-up approach to build a solution. The main idea is that in complex systems that evolve in dynamic environments, the entities

Sonia Suárez
Information and Communication Technologies,
University of A Coruña,
Campus de Elviña, s/n. 15071 A Corua, Spain
e-mail: ssuarez@udc.es

Paulo Leitao
Polytechnic Institute of Bragança, Department of Electrical Engineering,
Quinta Santa Apolónia, 1134. 4301-857 Bragança, Portugal
e-mail: pleitao@ipb.pt

Emmanuel Adam
UVHC, LAMIH (CNRS UMR 8201), F-59313 Valenciennes, France
e-mail: emmanuel.adam@univ-valenciennes.fr

J.B. Pérez et al. (Eds.): *Trends in Prac. Appl. of Agents & Multiagent Syst.*, AISC 221, pp. 103–111.
DOI: 10.1007/978-3-319-00563-8_13

have finite computation and communication capacities. Thus, the building of these complex systems arises from cooperation and coordination of the basic elements, following a kind recursive hierarchical architecture. Moreover, having a recursive architecture allows to use the same engine to build sub-entities and to delegate to these sub-entities the building of sub-sub parts of the system.

The holonic systems are typically a example of recursive systems.

According to A. Koestler [8], the term holon is used to refer to entities that behave autonomously - as a whole - but are not self-sufficient, as they behave as a part of a bigger whole, they need to cooperate with other holons. In this way, cooperating holons are able to establish a structure called holarchy. A holarchy can be defined as a system of self-regulated holons, which, depending on the circumstances, cooperate and/or collaborate in order to achieve a final goal. A holarchy can be conceived and handled as a single holon by abstracting its internal composition. Likewise, a holarchy may be made up of several holarchies. Accordingly, holarchies are simultaneously wholes and parts. Thus, holarchies follow a fractal structure, whose main characteristic is self-similarity, implying recursiveness or pattern-inside-of-pattern. In fact, the concept of fractal factories is used by [10] to propose a company composed of small components exhibiting autonomy and cooperation properties and the wholes-parties structure or recursive structure where the whole is more than the sum of the parts.

Holarchies are nowadays a well-known and wide-spread concept. In fact, they have been and still are being used in the manufacturing world to develop Holonic Manufacturing Systems (HMS) (see e.g. [9, 12]). But the use of holonic structures is not restricted to the manufacturing world and they are being extended to other fields (e.g. virtual enterprises [7]).

Irrespectively of the application domain, holonic systems are usually implemented using multi-agent systems (MAS) technologies, like in [9, 13] for example.

One of the main issue of the implementation of holonic systems remains their recursive architectures. Indeed, Calabrese et al. [3] concluded that holons natively show an architectural recursiveness that is a priori non defined in agents. This structural difference between holons and agents is the reason of this paper.

The work in this paper is motivated by the little effort dedicated in the literature to check the MAS technology adequacy for holonic implementation from the structural dimension, that is, to satisfy the inherent recursive structure of holonic systems. Most of the work is dedicated to justify the adequacy of MAS technology for holonic implementation based on the shared functional properties of autonomy and cooperation of agents and holons, but there is no discussion regarding these properties since they are natively properties of both holons and agents. This way, the paper analyses the concept of recursiveness and discusses how well and easily it can be implemented by using the most popular MAS technologies. As a result, a framework to compare MAS technologies (platforms) for the implementation of holonic structures is obtained that will guide the decisions of the designers and developers of this type of systems.

This paper is organized as follows. Section 2 discusses the concept of recursiveness and Section 3 presents the recursiveness implementation facilities with popular MAS frameworks. Section 4 exposes the synthesis and, lastly, section 5 presents the first conclusions of this work.

2 Holonic and MAS Recursiveness

Recursive systems are applications composed of similar entities. When recursiveness is discussed, functional or structural recursiveness can be considered. In *functional recursiveness*, definition of a recursive element contains a call to itself. *For example:* "$exp(n) = n \times exp(n-1)$, *with* $exp(0) = 1$". In *structural recursiveness*, the definition of an element contains links to elements of the same type. *For example:* "$chainedList = \{elt : int; nextElements : chainedList\}$"

In this work, we consider recursive system that refers to a system showing *structural recursiveness*. We take as example of recursive system a manufacturing company. It presents an organizational structure comprising different layers, as illustrated in Figure 1. It is typically a recursive system, each entity receives propose offers of services and receives orders that it takes in charge or not. Each entity decomposes an order into sub-orders proposed in its sub-parts.

Recursiveness is a widely accepted property for holarchies. Gerber et al. [5] say that manufacturing holons can be built recursively out of other holons and they give the name holonic agents to the recursive agents (i.e. agents consisting of sub-agents with the same inherent structure). They proposed tree generic possibilities for modeling holonic structures with MAS that differ in the degree of autonomy the sub-holons have and cover the spectrum from full sub-holon autonomy to a complete lack of autonomy. These possibilities are: (i) A holon as a federation of autonomous agents, where every sub-holon is fully autonomous and each super-holon is a new conceptual instantiation (e.g. HMS), (ii) several agents merge into one, that requires to terminate the participating sub-agents and to create a new agent as the union of the sub-agents (right way to model inanimate systems governed by physical laws like galaxies or organs), and (iii) a holon as a moderated group, where holons give up only part of their autonomy to the super-holon or head of the holon (e.g. virtual enterprises).

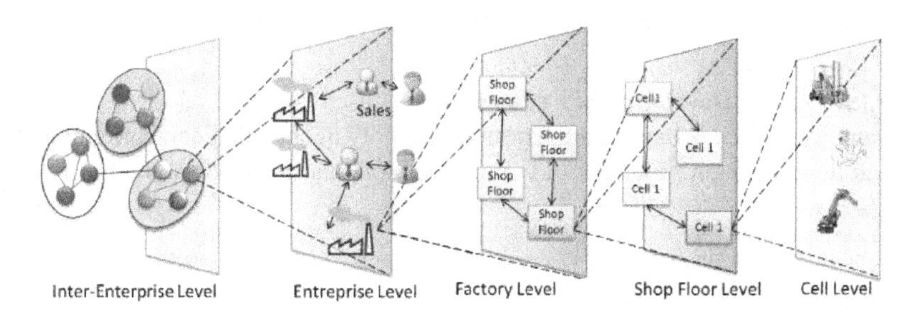

Fig. 1 Layered organization in manufacturing companies

As we focus in tis paper on animate systems (where entities are proactive), we consider in the following the first and third options. Regardless of which option is chosen, a holonic system is so a recursively enumerable set of holons.

As it was said previously, holonic systems (i.e. holarchies) have two main functional properties, namely autonomy and cooperation (as they are holons) and the structural property of recursiveness. The implementation of holonic systems by using the MAS technologies allows the truly implementation of the holonic functional properties, since both holons and agents are autonomous and not self-sufficient. Nevertheless, the truly implementation of the structural property of recursiveness is not an easy and clear task as agents are not natively recursive. Having this in mind, this section discusses the possibilities to implement that structural characteristic with the mostly used MAS development frameworks. In particular, it will be discussed Jade (Java Agent DEvelopment framework [2]) and Jack[1] and its extension JackTeams[2] as they are the most used for the implementation of MAS. Janus platform [4], MadKit (MultiAgent Development KIT [6]) and its extension Madkit-GroupExtension[3] will also be briefly discussed. In all cases, the attention is devoted to recursiveness.

For each framework, we study how we can define *groups of agents*; and how it is possible *to develop a recursive structure* and what kind of recursiveness is generated. We study also the notion of *roles* allowing to identify if it is possible for groups of agents to play roles in other groups of agents.

Thus, when implementing recursive structures (i.e. holarchies) it is important to bear in mind that: (i) holons are independent from holarchies, (ii) a holon exists without belonging to any holarchy, (iii) a holon can dynamically join a holarchy without stop being independent and (iv) a holon can join more than one holarchy at the same time.

In this paper, we propose a small study to compare the implementation of recursiveness with MAS development frameworks applied to an example in the field of the HMS and we try to implement the example following the first and the third possibilities described previously. We consider a *Shop Floor* that receives orders to assemble products; this shop floor is composed of an *Assembly Cell* and a *Transport Cell*. The assembly cell is composed of four *Resources* (workstations); the transport cell is composed of three *resources* (AGV for AutoGuided Vehicle).

Jade (Java Agent Development Framework)

The framework Jade is based on the notion of empty agents, that can communicate between them through protocols of communication, and in which some behaviour

[1] Jack platform web site is : `http://aosgrp.com/products/jack/`

[2] JackTeams extension web site is : `http://aosgrp.com/products/jackteams/`

[3] Madkit Group Extension web site is : `http://madkitgroupextension.free.fr/`

can be added (initially, or at run time). Implementing a recursive system (holarchy) with this framework implies that this holarchy extends the Jade class Agent. It is possible by defining a particular class ('*JadeHolon*') that extends the Jade class Agent and that defines the notion of: supervisors (set of the super holons AIDs[4]); assistants (set of sub-holons AIDs); and a list of holonic behaviours guiding theirs actions relatively to the environment, and to theirs acquaintances.

In [1], the notion of functional roles is used to develop holonic systems. A list of roles is associated to a '*JadeHolon*'. A role represents here a set of behaviours that identify a type of agent. So JADE allows to define a federation of agents, and, if roles are implemented, it allows to define set of agents playing the same roles.

The **Janus platform** [4] is a FIPA compliant platform, in some points similar to JADE, but enhanced with the notion of functional roles and that uses the concept of groups. When a Janus agent could implement a Holon predefined interface, it can contains a list of groups where are situated other agents. Thus the notion of recursiveness developed like a federation of agents is also possible in Janus.

To apply JADE to the example (we choose to present an application of Jade because it is more used at the moment than Janus), the following tree classes were implemented by extending the '*JadeHolon*' class (that extends the Jade class '*Agent*'): '*Resource*', '*AssemblyCell*' and '*TransportCell*'. The first one represents a single holon, and the second and the third classes represent holarchies, i.e, holons with component parties. Four roles have been implemented: '*TransportResourceRole*', '*AssemblyResourceRole*', '*AssemblyCellRole*' and '*TransportCellRole*'.

To have an implementation compatible with the first modelling possibility, an instance of a '*Resource*' contains in its '*supervisors*' attribute the AIDs of the instances of the '*AssemblyCell*' and '*TransportCell*' to which it is linked; and these instances store in their '*assistants*' attribute the AID of the of the associated instance of'*Resource*'.

To have an implementation compatible with the third modelling possibility, only the definitions of the roles link the agents that play the '*AssemblyCellRole*' and '*TransportCellRole*' to agents that play the '*xxxResourceRole*'.

Jack and Jack Teams

The philosophy of Jack is comparable to that of JADE. The main difference is that Jack works with Belief-Desire-Intention (BDI) agents architecture.

Particular attention needs to be paid to its extension JackTeams. The main structural classes in Jack Teams are '*Team*' and '*Role*'. Every agent (single or composed) is an instance of the class '*Team*' (extension of Jack class '*Agent*').

A team instance defines the roles that it can play in other teams and the own roles that are required to be played by other teams. These sub-teams have their own roles, with the same structure. As a result, a nested structure of teams and roles is created. For team structure formation, it is necessary to indicate which teams are capable of perform which roles in the structure (the structure can be modified in runtime).

[4] An agent AID is its Identifier inside the Jade platform.

This way, a holarchy is defined as a team instance with a set of roles. Each role is played by others holons/holarchies; that is, others team instances (in JackTeams everything is a team instance).

Schematically, for the manufacturing example a '*Team*' class '*ShopFloor*' represents the entire holarchy. This class requires the roles '*AssemblyCellRole*' and '*Transport-CellRole*'.
In addition, the teams '*AssemblyCell*' and '*TransportCell*' require the role '*Resource-Role*'. Specifically '*AssemblyCell*' requires four players for the role and '*Transport-Cell*' requires three. The Team class '*Resource*' performs the roles '*AssemblyCellRole*' and '*TransportCellRole*' without requiring any new role.
This is an implementation compatible with the first and the third modelling possibilities, according to whether a role head is defined or not in the team instances.

Madkit, MadKitGroupExtension
The main structural classes in *MadKit* are '*Agent*', '*Group*' and '*Role*'. An agent is only specified as an active communicating entity which plays roles within groups. The agent designer is responsible for choosing the most appropriate agent model as internal architecture. Groups are defined as atomic sets of agent aggregation. Each agent is part of one or more groups. A group may be founded by any agent, and an agent may request the admission to any group. Lastly, the role is an abstract representation of an agent function, service or identification within a group. Indeed, the working mechanism of an agent is stored inside its class, not in its roles. The notion of roles for an agent allows it to be identified by other agents. Each agent can handle multiple roles, and each role handled by an agent is local to a group.

MadKitGroupExtension allows the creation of advanced group representations. In *MadKit* a group contains a set of agents playing roles. In *MadKitGroupExtension*, a group can be contained in others groups. Alike, a group can contain subgroups.

Continuing with HMS example, some details about *Madkit* and, specifically, *Madkit-GroupExtension* are given. Three '*Group*' classes ('*ShopFloor*', '*AssemblyCell*' and '*TransportCell*') are defined. The last two classes are subgroups of the first one.
At the lowest level the '*Agent*' class '*Resource*' appears. Thus, the group '*Assembly-Cell*' includes four '*Resource*' instances and the group '*TransportCell*' includes three instances, playing different roles. Note that '*Resource*' is the unique functional unit. The rest of the classes are structural units (similar to the list of folders and subfolders in a directory).

3 Synthesis

In the Jade implementation, the '*AssemblyCell*' and '*TransportCell*' instances know the AIDs of their components and the instances of '*Resource*' know their holarchies (i.e. their super-component). At the same time, instances of '*Resource*' can join other holarchies and/or they can work autonomously. However, these component agents themselves are not really structural components of a bigger whole. This is a structure quite similar to the recursive structure (like in holonic systems), but it need to define a specific '*HolonAgent*' class and its associated behaviour like developed in [1] for example.

In Madkit and, more specifically, in MadkitGroupExtension the groups are like folders or containers with agents, but a group can not be seen from the outside as an agent (i.e. as a new holon). In other words, a group can not play a role in other group. This is because the functional unit is not the group but the agent; the group is only a representation artefact. Thus, it does not represent a recursive structure like in holarchies.

In Janus, it is possible to use both functional roles, and the notions of roles like in Madkit. But in Janus, agents contains list of roles that regroup other agents.

Jack Teams allows the creation of a nested structure of teams and roles. This way, a holarchy is defined as a team instance with a set of roles. Each role is played by other holon/holarchy; that is, other team instance. That is to say, single holons and holarchies are not instances of different classes, all of them are team instances. If we want to know if a given team instance is a single holon or a holarchy we have to look inside the object and find if it both requires and performs roles or if it only performs roles. In the first case it is a holarchy (it will include other holons playing the roles) and in the second case it is a single holon (it will not include other holon). From an external point of view the entire structure (i.e. any team, single or composed) is seen as an unique holon. This is absolutely in accordance with Koestler's ideas about the whole-part structure or recursive structure of holarchies. Thus, both the teams and the single holons are functional entities behaving as a whole and/or as a part of a bigger whole. Thus, with JackTeams the resulting system is inherently and naturally recursive as this is a consequence of the classes themselves. Nevertheless, with Jack and JADE, developers must try to simulate this recursive structure explicitly and ad-hoc, as these tools are not equipped with the right mechanisms to modelling recursive structures.

Table 1 presents a short summary of our evaluation, focusing on the frameworks Jade/Jack, JackTeams, MadKitGroupExtension, and Janus. The term 'Recursivity implementation facilities' resumes the way to develop a recursive system with each framework, and 'Holonic recursivity' refers to the pertinence of the framework to implement recursiveness as described in the holonic paradigm.

Table 1 Summary of MAS platforms adequation to recursive systems

Platform	Agent sets	Roles	Recursivity implementation facilities	Recursivity Model	Holonic recursivity
JADE, JACK	Services	Functional roles (to be developed)	Need specific classes	Federation of autonomous agents	Implementation ad-hoc
JACKTEAMS	Teams, Roles	Functional roles	Teams defined by roles played by teams ...	Federation of autonomous agents and by moderated groups	Natively, by the classes Team and Role
MADKITGroup	Groups, Roles	Artifact (that just helps to group agents)	Groups contain groups	By moderated groups	Not available (a group can not play a role in other group)
JANUS	Services	Functional roles	Agents contain groups of agents	Federation of autonomous agents	Natively, by the classes Group and Role

4 Conclusion

In order to manage and develop complex and large systems, it is often needed or recommended to build recursive systems, in order to easily define and deploy a large number of entities on a network. As recursiveness is an elemental structural property of the holonic systems, often used to represent complex manufacturing systems, we tried to find the best MAS development framework able to support holonic systems development, and specifically the notion of recursiveness. For this, we compare in this paper the most used MAS development frameworks and add-ons of them regarding the notion of recursiveness; but the evaluation strategy is applicable to any new framework. We think that the evaluation results will help designers and developers of holonic systems to select the appropriate MAS development framework for their own needs. Thus, we think that this first study will allow the responsible selection of MAS development frameworks useful for the implementation of these type of recursive systems.

Acknowledgements. The authors would like to thank Javier Andrade (University of A Coruña, Spain) for his support on holonic theoretical aspects.

References

1. Adam, E., Mandiau, R.: Flexible roles in a holonic multi-agent system. In: Mařík, V., Vyatkin, V., Colombo, A.W. (eds.) HoloMAS 2007. LNCS (LNAI), vol. 4659, pp. 59–70. Springer, Heidelberg (2007)
2. Bellifemine, F., Caire, G., Poggi, A., Rimassa, G.: Jade - a white paper. Tech. rep., TiLab (2003), `http://jade.tilab.com/papers/WhitePaperJADEEXP.pdf`
3. Calabrese, M., Piuri, V., Di Lecce, V.: Holonic systems as software paradigms for industrial automation and environmental monitoring. In: Intelligent Agent (IA), IEEE Symposium on Intelligent Agent, Paris, pp. 1–8 (2011)
4. Gaud, N., Galland, S., Hilaire, V., Koukam, A.: An Organisational Platform for Holonic and Multiagent Systems. In: Hindriks, K.V., Pokahr, A., Sardina, S. (eds.) ProMAS 2008. LNCS, vol. 5442, pp. 104–119. Springer, Heidelberg (2009)

5. Gerber, C., Siekmann, J., Vierke, G.: Holonic multi-agent systems. Tech. rep., DFKI (1999)
6. Gutknecht, O., Ferber, J., Michel, F.: Integrating tools and infrastructures for generic multi-agent systems. In: Proceedings of the Fifth International Conference on Autonomous Agents, AGENTS 2001, pp. 441–448. ACM, New York (2001)
7. Huang, B., Gou, H., Liu, W., Li, Y., Xie, M.: A framework for virtual enterprise control with the holonic manufacturing paradigm. Computers in Industry 49(3), 299–310 (2002)
8. Koestler, A.: The Ghost in the Machine. Arkana Books (1969)
9. Leitao, P., Restivo, F.: Adacor: A holonic architecture for agile and adaptive manufacturing control. Computers in Industry 57(2), 121–130 (2006)
10. Warnecke, H.J., Owen, T.: The Fractal Company: A Revolution In Corporate Culture, vol. 11. Springer, Berlin (1993)
11. Simon, H.A.: The Sciences of the Artificial, 3rd revised edn. MIT Press (1996)
12. Valckenaers, P., Brussel, H.V., Wyns, J., Bongaerts, L., Peeters, P.: Designing holonic manufacturing systems. Robotics and Computer-Integrated Manufacturing 14, 455–464 (1998)
13. Zambrano Rey, G., Pach, C., Aissani, N., Bekrar, A., Berger, T., Trentesaux, D.: The control of myopic behavior in semi-heterarchical production systems: A holonic framework. Engineering Applications of Artificial Intelligence 26(2), 800–817 (2013)

Intelligent Energy Management System for the Optimization of Power Consumption

Vicente Botón-Fernández, Máximo Pérez Romero,
Adolfo Lozano-Tello, and Enrique Romero Cadaval

Abstract. This paper describes a Smart Storage System capable of managing energy and smart home devices to optimize the local power consumption of a house. The proposed model consists of two main systems; the Local Energy Management Unit (LEMU) and the Central Energy Management and Intelligent System (CEMIS). On the one hand, the LEMU is able to maintain the power consumption under a maximum reference value and to switch on/off the devices by using domotic protocols. On the other hand, the CEMIS receives operation data remotely from several devices, analyzes them using intelligent techniques and determines the best operation strategy for each LEMU, communicating the operation references back to them.

1 Introduction

Nowadays, the availability of having Energy Storage Systems (ESS) [1, 2] is a key factor for establishing new operation strategies in electric distribution grids. Such systems would help to integrate distributed generation systems based in renewable energies into houses and also would contribute to get some control on the load profile, and so achieving a better overall performance of the distribution grid.

This kind of system is typically associated with distributed generation plants but an ESS could also be used in the absence of an energy generation system, with the aim of smoothing the load curve or even controlling the consumption of

Vicente Botón-Fernández · Adolfo Lozano-Tello
Quercus Software Engineering Group – University of Extremadura
Escuela Politécnica, Campus Universitario s/n, 10071, Cáceres, Spain
e-mail: {vboton,alozano}@unex.es

Máximo Pérez Romero · Enrique Romero Cadaval
Power Electrical and Electronics R+D Group – University of Extremadura
Avda. De Elvas s/n, 06006, Badajoz, Spain
e-mail: {mperez,eromero}@unex.es

J.B. Pérez et al. (Eds.): *Trends in Prac. Appl. of Agents & Multiagent Syst.*, AISC 221, pp. 113–120.
DOI: 10.1007/978-3-319-00563-8_14

energy depending on its price-by-hour [3]. An important feature that these systems need to possess is the ability to adapt themselves to the users' consumption profile and have the versatility to support decision-making in different situations. Identifying consumption patterns within a sequence of events in order to forecast future peaks on the load curve, can help us achieve energy optimization.

An appropriate way to understand the load profile of a building and infer new information can be through the use of ontologies and SWRL (Semantic Web Rule Language) [4] to classify the types of appliances and sensors and their operation.

In this paper an Intelligent Energy Management System (INTELEM) is presented. This system comprises LEMUs, which are located in houses, and a central unit that receives data from a group of LEMUs via TCP/IP connections. This central unit analyses the information using behavioral algorithms in order to design a strategy that can optimize the load curves. We introduce the grounds we will take as a starting point for the development of those algorithms.

2 Data Mining for Forecasting Power Consumption

The development of data mining techniques to forecast usual behavior and provide feedback in context constitutes a key factor to empower users to take control over residential power consumption. There are several factors that drive household power consumption behavior, such as energy-related attitudes, socio-demographic factors, energy prices, etc. van Raaij and Verhallen [5] suggest that habits can become alternative predictors of power consumption, because habits may resist the cognitive and financial drive and still prevail over rational alternatives.

In this sense, Joana M. Abreu et al. [6] propose a methodology which demonstrates that it is possible to use pattern recognition methodologies to find habitual power consumption behavior given the intrinsic characteristics of the users.

The proposed work aims to achieve similar objectives but with a different approach. It presents the design of an intelligent system that will recognize power consumption patterns to automate the grid behavior in an efficient way, providing electricity supply when needed and saving as much energy as possible. The behavioral algorithm will take as starting point the analysis of temporal energy consumption data and the use of the building resources.

3 INTELEM

The Intelligent Energy Management System presented in this paper, is built upon an existing project called IntelliDomo[1] [4]. It is a smart system able to control the devices of a home automation system automatically and in real time using SWRL rules and learning algorithms.

---

[1] `http://www.intellidomo.es`

Figure 1 shows INTELEM architecture. As it can be seen, the CEMIS can take control of several LEMUs and is composed by two main modules: the Information Management Module (IMM) and the Energy Optimization Module (EOM). It's important to remark that these LEMUs are distributed over different houses or residential buildings whereas the CEMIS is located at a central server, using a conventional connection via TCP/IP to exchange information with all these LEMUs.

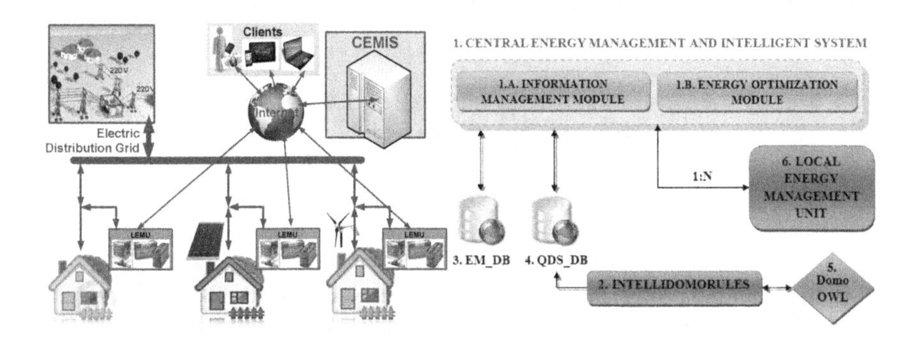

Fig. 1 This figure shows the components of INTELEM architecture and their relations

On the one hand, the CEMIS receives data from a group of LEMUs and stores it into a central database called EM_DB. On the other hand, the CEMIS will analyze the data logs with a behavioral algorithm in order to find energy consumption patterns and use this information to generate production rules which can optimize the use of the grid and save energy. Then these rules are stored into the QDS_DB database and managed through the IntelliDomoRules tool. Finally, once the rules are fired, the corresponding changes are transmitted to the affected LEMUs.

3.1 Local Energy Management Unit

The main function of this device is to control the energy consumption of a house or residential building. A LEMU is able to limit the power consumption peaks, which can take place due to start-up transients or high power devices operating during short-time intervals, by supplying extra energy from the ESS. As a result, the user could contract with the electric power company to reduce the level of contracted power that they would normally have without this device. As this contracted power is usually included as a fixed cost in the electricity invoice, the user could have direct benefits.

Apart from that, a LEMU can also decouple the consumed energy from the energy demanded to the grid. As far as time-of-use (TOU) rates are concerned, it would be possible to charge the ESS from the grid during the off-peak hours and then supply its energy to the devices during the peak hours. This way, a LEMU could work as a constant power load, by using the ESS to put the difference of energy between the energy demanded from the grid and the current level of energy consumed by the devices into the house. Figure 2 describes its main structure.

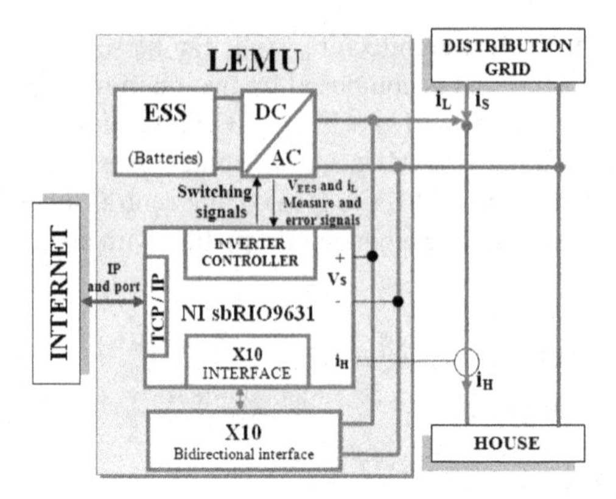

Fig. 2 This figure shows the structure of the Local Energy Management Unit

Moreover, a LEMU is able to switch on/off the devices by using the X10 protocol to maintain the power consumption under an established limit, and it can also receive the X10 events sent by X10 sensors such as motion sensors.

Finally, the LEMU has been designed to be integrated into an energy management network, and it has the possibility of being connected to a central system in order to send local data, such as electrical magnitudes, the state of charge of its ESS or X10 events. The LEMU receives operation references from the central system such as power consumption limits, or some X10 commands.

3.2 Central Energy Management and Intelligent System

This central unit is in charge of coordinating the group of LEMUs which are distributed over the grid. It's designed to be hosted in a remote server and its main objective is to efficiently manage the power consumption of a house through data mining algorithms. Basically, the LEMUs gather power consumption data through the sensors that are connected to the grid, sending that information to the CEMIS. Then the CEMIS can analyze it and use it to send control signals back to the LEMUs to coordinate their behavior and save energy. As stated in Section 3, the CEMIS has two modules to achieve those objectives: the IMM and the EOM.

3.2.1 Information Management Module

This module is continuously reading and storing into a database the information provided by the LEMUs. The data sent by the LEMUs can be values of electrical magnitudes, X10 events or the state of charge of an ESS. The electrical magnitudes are often sent every 5 seconds, whereas the other values are sent when a certain event takes place (e.g. an X10 sensor detects motion). In order to better identify each type of information, it has been defined three different message format, all of them with the following structure: header field plus data field.

The X10 message format is as follows: header = "X10", data = {house code, unit code, value}. The data field uses one character to represent the house code (A-P), two characters to describe the unit code (1-16) and three more characters to identify the value. The ESS message format has the following fields: header = "BAT", data = state of charge (%). Finally, the power consumption message format uses the data field to store the value of electrical magnitudes (kW) and its header is "CON".

Otherwise, these messages are transmitted through a socket between each LEMU and the CEMIS. In this case, it's the LEMU who takes the client role and the one who initiates the communication session with the CEMIS, which is waiting for incoming requests. Once the socket is established, the LEMU will send data packets continuously, and then the IMM will process them and save the information into the EM_DB. This way, the database keeps a log of everything that happens inside the houses. Obviously, after each packet reception the IMM will response with an acknowledgement message.

3.2.2 Energy Optimization Module

The main objective of this module, whose algorithms are still in development, is to coordinate a group of LEMUs, optimizing the use of the grid and saving energy. To that end, this module incorporates data mining algorithms with the purpose of controlling the ESS charging time. These algorithms aim to discover the resident power consumption patterns and try to have the ESS prepared for those moments where the consumption is considerably high. With this information, the system could predict future demands on the grid and manage them with the best strategy. Figure 3 sums up the operation of this module.

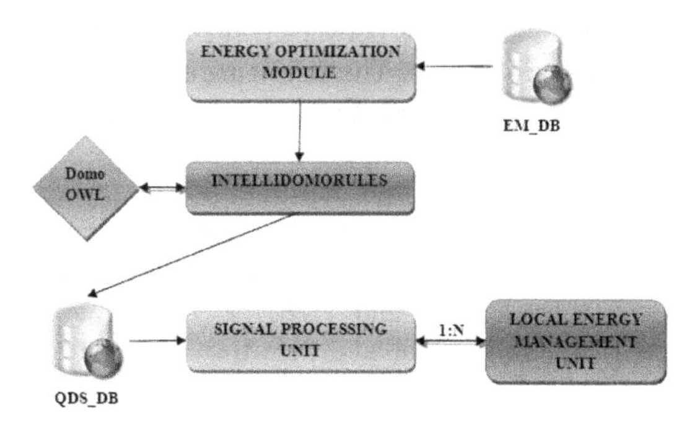

Fig. 3 This figure shows how the Energy Optimization Module works

First of all, the EOM reads the log entries from the EM_DB database. These entries are composed of three fields; power consumption value, timestamp of the event and IP of the associated LEMU. Then the EOM selects a specific LEMU and sends the corresponding entries to the data mining algorithm (see section 4). After that, the algorithm will analyze this data set and look for frequent peaks in the load curves, designing a strategy to coordinate the LEMUs and optimize the use of the resources. The way to carry out this strategy involves generating SWRL rules that automate the corresponding actions. As a result, the EOM will send the outcomes of the data mining process to the IntelliDomoRules tool. This one can build the rules and save them into the ontology and the database.

The Signal Processing Unit (SPU) monitors the rules of the QDS_DB database. When a rule affecting a LEMU is fired, it sends a control signal to the associated LEMU. There are two types of control signal: one to control the charge of the ESS and another one to manage X10 devices. The control signals for X10 devices follow the same format as that used in the IMM-LEMU communication, but now these signals can remotely modify the physical state of an X10 device. This type of signal is usually used to switch low-priority X10 devices off when the consumption curve reaches a peak and there is not enough power supply in the ESS. Otherwise, the first type of signals has a header value of "BAT" and it uses the data field to store the ESS command, which would be "on" if the EOM wants to start the charge of the ESS or "off" if it wants it to stop the charge. To communicate the SPU with a LEMU it's necessary to create another socket. However, this time it's the CEMIS who takes the role of the client and each LEMU will work as a server, listening for new requests. The SPU sends a control signal and then receives an acknowledgement response from the LEMU. Once the control signal has been transmitted, the socket is closed and the LEMU waits again for new requests.

4 Data Mining Algorithm

Data collected in a smart environment consist of readings by sensors and resident manual manipulation of devices. Each event from the data set is considered an action; the goal is to find relationships between them to better explain the circumstances that cause consumption peaks. The following algorithm is inspired in both Apriori's Algorithm [7] and the Episode Discovery Algorithm [8].

Given a set E of events types, an action is a pair (A, t), where $A \in E$ is an event type and t is the timestamp of the action. Each event type contains two attributes: the data source and the value of the source. An action sequence **s** on E is a triple (s, T_s, T_e), where $s = \{(A_1, t_1), (A_2, t_2), \ldots, (A_n, t_n)\}$ is an ordered sequence of actions such that $A_i \in E\ \forall\ i = 1, \ldots, n$, and $t_i \leq t_{i+1}\ \forall\ i = 1, \ldots, n - 1$. Further on, T_s and T_e are timestamps, where T_s is called the starting time and T_e the ending time, and $T_s \leq t_i < T_e\ \forall\ i = 1, \ldots, n$.

The main objective of this analysis of sequences is to find all frequent behavior patterns from a class of patterns with a suitable error rate. We are interested in those classes of patterns whose events involve power consumption peaks. The user defines the threshold that determines if a power consumption value can be considered as a peak. This threshold is referred as *minimum consumption min_cp*. Otherwise the actions of a frequent pattern must take place close enough in time. The user defines how close is close enough by giving the width of the *action range* within which the pattern must occur. An *action range* is defined as a slice of an action sequence. Formally, an *action range* on an action sequence $\mathbf{s} = (s, T_s, T_e)$ is also a sequence of actions $\mathbf{r} = (r, t_s, t_e)$, where $t_s < T_e$ and $t_e > T_s$, and r consists on those pairs (A, t) from s where $t_s \leq t < t_e$. The length of the *action range* is $t_e - t_s$ and it is denoted $length(\mathbf{r})$. Given a data set, the user provides an integer *len* and then the algorithm looks for frequent patterns using an *action range* **r** such that $length(\mathbf{r}) = len$. To be considered interesting, the algorithm must focus on those *action ranges* containing a pair (A', t), where $A' \in E$ is a type of event that represents a power consumption peak.

The periodicity α represents the regularity of occurrence of patterns. In order to compute the frequency of a class of pattern, this algorithm considers several types of periodicity: daily, weekly, monthly, annual...

A pattern p is described as a frequent and well-defined collection of actions occurring together. It consists of a triple $(N, \leq, g)$ where N is a set of nodes and $g : N \rightarrow E$ is a mapping associating each node with an event type. The events in $g(N)$ have to occur in the order described by $\leq$. The size of p, denoted $|p|$, is $|N|$. The nodes of the pattern need to have corresponding actions in the sequence such that the event types are the same, the order of the pattern is respected and it happens in the given *action range*.

A group of actions is considered frequent when its frequency reaches a certain threshold referred as *minimum support min_sp*. It is the user who specifies how many times a collection of actions has to occur in a period of time to be considered frequent. Given an action sequence **s**, a periodicity α and an *action range* length *len*, the frequency of a pattern p in **s** is

$$f(p,s,\alpha,len) = \frac{\left|\left\{r \in s \mid p \text{ occurs in } r \wedge length(r) = len\right\}\right|}{h(T_s,T_e,\alpha)} \quad (1)$$

where $h(T_s, T_e, \alpha)$ is a function that counts the number of days of the time interval $[T_s, T_e)$ according to the periodicity α, e.g. if α represents a weekly periodicity, then for each week in $[T_s, T_e)$ it only counts one day. Therefore, a group of actions is a frequent pattern if $f \geq min\_sp$. As it was mentioned before, our system is interested in discovering all frequent patterns from a specific class of patterns, which are those related with power consumption peaks.

Once the patterns are learned, the CEMIS could use them to obtain production rules that describe relations between different actions and forecast future consumption peaks. The conversion of these patterns into rules was described in [4].

5 Conclusions and Future Works

There are many opportunities to use INTELEM in Smart Houses or Buildings to achieve economic benefits and an optimized use of the distribution grid. These opportunities increase if the house or building has a distributed generation system, which is another factor that must be taking into account when defining the operation strategy. In this sense, this work presents the bases of an intelligent energy management system that aims to achieve these objectives, describing the relations and the main features and functions of each component.

At the present time, INTELEM can control several LEMUS to optimize the power consumption and manage smart home devices remotely. In our ongoing work, we're building data mining algorithms upon the grounds presented here in order to coordinate the behavior of a set of LEMUS in an efficient way, so that we can optimize the use of the distribution grid in different houses at the same time while saving energy altogether. We will test these behavioral algorithms in different scenarios to better understand their strengths and weaknesses.

Acknowledgments. This work has been developed under support of TIN2011-27340 (Spanish Contract MIGRARIA), Telefónica Chair of Extremadura University and Junta de Extremadura.

References

1. Pieper, C., Rubel, H.: Revisiting Energy Storage: There Is a Business Case. The Boston Consulting Group (2011)
2. Guerrero, M., Romero, E., Barrero, F., Milanés, M., González, E.: Supercapacitors: Alternative Energy Storage Systems. In: PRZEGLĄD ELEKTROTECHNICZNY (2009)
3. Auväärt, A., Rosin, A., Belonogova, N., Lebedev, D.: NoordPoolSpot Price Pattern Analysis for Households Energy Management. In: CPE-2011 Workshop, Tallinn, Estonia, pp. 103–106 (2011)
4. Lozano-Tello, A., Botón-Fernández, V.: Analysis of Sequential Events for the Recognition of Human Behavior Patterns in Home Automation Systems. In: DCAI-2012, Salamanca, Spain, March 28-30, pp. 511–518 (2012)
5. van Raaij, F., Verhallen, T.: A Behavioral Model of Residential Energy Use. Journal of Economic Psychology 3(1), 39–63 (1983)
6. Abreu, J.M., Câmara Pereira, F., Ferrão, P.: Using pattern recognition to identify habitual behavior in residential electricity consumption. Energy and Buildings 49, 479–487 (2012)
7. Agrawal, R., Srikant, R.: Mining Sequential Patterns. In: ICDE-1995, Taipei, Taiwan, March 6-10, pp. 3–14 (1995)
8. Youngblood, G.M., Cook, D.J.: Data Mining for Hierarchical Model Creation. IEEE Systems, Man, and Cybernetics, Part C: Applications and Reviews 37(4), 561–572 (2007)

Find It – An Assistant Home Agent

Ângelo Costa, Ester Martinez-Martin, Angel P. del Pobil,
Ricardo Simoes, and Paulo Novais

Abstract. Cognitive impaired population face with innumerable problems in their daily life. Surprisingly, they are not provided with any help to perform those tasks for which they have difficulties. As a consequence, it is necessary to develop systems that allow those people to live independently and autonomously. Living in a technological era, people could take advantage of the available technology, being provided with some solutions to their needs. This paper presents a platform that assists users with remembering where their possessions are. Mainly, an object recognition process together with an intelligent scheduling applications are integrated in an Ambient Assisted Living (AAL) environment.

1 Introduction

Countries and societies evolve at a high rate leading to new social problems to be dealt with. One of these problems is related to the population undergrowth in developed countries. The reason for that undergrowth lies on the fact that birth rate

Ângelo Costa · Paulo Novais
CCTC - Computer Science and Technology Center,
Department of Informatics, University of Minho, Braga, Portugal
e-mail: {acosta,pjon}@di.uminho.pt

Ester Martinez-Martin · Angel P. del Pobil
Robotic Intelligence Lab, Engineering and Computer Science Department,
Jaume-I University, Castellón, Spain
e-mail: {emartine,pobil}@uji.es

Ricardo Simoes
Institute of Polymers and Composites IPC/I3N, University of Minho,
Guimarães, Portugal, Life and Health Sciences Research Institute (ICVS),
School of Health Sciences, University of Minho, Campus de Gualtar,
4710-057 Braga, Portugal, Polytechnic Institute of Cávado and Ave,
Barcelos, Portugal
e-mail: rsimoes@dep.uminho.pt

J.B. Pérez et al. (Eds.): *Trends in Prac. Appl. of Agents & Multiagent Syst.*, AISC 221, pp. 121–128.
DOI: 10.1007/978-3-319-00563-8_15

is continuously decreasing, while life expectancy is increasing. For that reason, from the current society model, two problems arise: medical response and economic stability.

Speaking in terms of cognitive problems, the increase in elderly people implies that the common problems affecting them, grow in incidence. Regarding economy, country production has a severe setback: population cannot get the necessary income to support everyone. Therefore, expensive services are beyond most of the population.

In addition, the cognitive problems are very crippling in terms of independence. What is more, these problems can lead to other health problems. So, for instance, it is common to lose balance and, consequently, fall down. This fact could result in broken bones and other diseases that can affect severely the elderly life, resulting even in death (in extreme cases). In terms of daily life, a person needs constant attention when the cognitive impairment is severe. That attention can be provided by a personal caregiver or by living at a specialized institution. Consequently, these people need little help in their everyday activities, allowing them to live alone, as well as more cognitive aids like refreshing their memories, are required.

Elderly population does still not have all the help they need. For that, designing an agent aiding people to remember where their possessions are at home is imperative. This paper addresses the general directions for the design and implementation of an intelligent agent aimed at this issue. We present a modular architecture such that, from visual information gathered at home by a set of cameras, the agent will provide the user with the location of their most common possessions like their keys.

2 General Directions of AAL-Related Research

Ambient Intelligence can be defined as the research focused on sensitive and responsive electronic environments, involving the user. That is, it is aimed to provide people with technological solutions to help them from perceiving and analyzing their environment. As part of the AmI area, we can find the Ambient Assisting Living (AAL). In particular, the AmI covers all the electronic devices, while the AAL is focused on any user with disabilities and his/her home.

Despite the AAL progress, it is still a diffuse area because each project is aimed at solving a particular task, but not at being part of a whole integrated environment of services. Consequently, the lack of cooperation between the developed projects results in a group of services that do not communicate each other, loosing precious features that could outcome from projects' fusion. Furthermore, without well-established standards, it is nearly impossible to integrate different projects into a more complete, practical system. On this matter, Norgall et al. [20] analyzed the latest developments and standards on this topic. In this study, they proposed some advices on development as well as future global standards. Among them, it should be pointed out the architecture, communication and physical implementation. The AAL4ALL project is a pioneer on this topic. It is a consortium of technological and medical partners that work on producing and implementing complete AAL platforms. These platforms can be used in user's home, in a hospital or at a day-centre [26]. The multi-agent system is able to connect different developments in the same platform, integrating them in a unified service. So, information can be exchanged and linked resulting in more complex and sophisticated services [1].

From this starting point and keeping in mind the user's needs, we propose a complete, modular system to assist a user in remembering his/her activities as well as all the required possessions before leaving his/her home (e.g. home keys) and where they are, if necessary. With that aim, we present a multi-agent home platform where a vision module for object recognition and location is integrated. So, the implemented multi-agent platform is presented in Section 3, while the vision module is described in Section 4. Finally, some conclusions and future work is discussed in Section 5.

3 Multi-Agent Home Platform

The multi-agent home platform, which consists of a HAN (Home Area Network) and BAN (Body Area Network), monitors and helps a person to perform their daily tasks. Every module contains, at least, two agents: one for action control; and another for communication. Note that, in a further stage, some sensors could be attached to user clothes for capturing their body signs.

The iGenda project is one part of this system [7, 8]. It offers a mobile system (on a Smartphone), in terms of interface, to the user, that is, it provides an intelligent schedule system and an appealing interface. The scheduling system handles new events sent from different users or platforms, and leisure events. In addition, a profiling system has been integrated to improve the decision module. The module is connected to the server framework for sending and receiving information about the operations and actions occurring within the platform, as shown in Fig. 1.

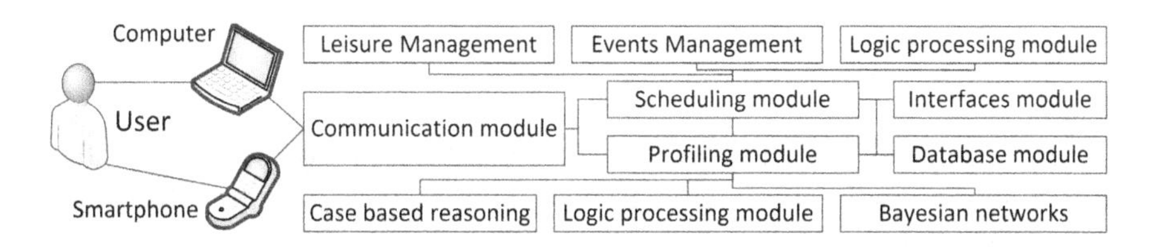

Fig. 1 iGenda modules and systems overview

Since the mobile platform is implemented over the Android operating system [22], it has full support to the Google services, including calendar, email and chat. With the aim of a successful performance, the system must be permanently active, even when there is no connection with the server. Therefore, the system is composed of different agents managing each feature. Mainly, these agents are divided into communications, logic decisions, interface bridging and database management. Among all these features, the profiling system is the most important to this project merging action.

The profiling system consists of an agent integrated in the scheduling module. This agent is in charge of controlling the module learning process by keeping all the information relative to user interactions. That is, all the received user edits and sends of events are registered in an event history [13]. Two types of learning systems are used to emulate the user preferences: Case Based Reasoning (CBR) and Bayesian Networks (BN) [17, 27].

With the aim of making the concept clearer, think about the traveling salesman problem. It has the same implementation foundations, evolving to a multi-level weighted graph. Thereon, heuristic compounds allow to creating probabilistic computational tree for each event. In terms of tree limitation or pruning, a trust value of 80% is used to constrain the fractal evolution. That is, the sum of three or less tree vectors must be above 80%, where the vectors refer to the number of times an event has been used to modulate the user's decision. Another highlighting feature of the profiling agent is keeping the overall user's preferences since this information could be useful for other modules. For instance, tracking the common user's objects and their location makes easier to provide information about them to the user. Therefore, when an object is required for performing an activity, the object recognition system could be used to remind where it is.

The goal is to open a path to connect other projects with the iGenda. That is, opening a way to share the information that both projects have. As previously stated, our perspective relays on joining developments, not on building a new project. Due to the nature of the iGenda project and its aim, an object recognition feature is the best solution to aid the user locating objects. Thus, it can result in the following scenario: if the user needs to exit its home he/she has to take the house keys, thus, the platform can warn the user that the keys are needed and, at the same time, show where they are.

4 Object Recognition and Location

Object recognition is one of the most important problems in Computer Vision. For decades, researchers have introduced many algorithms in the hope of building systems with the ability of visually perceiving their surrounding environment. In recent years, with the advances in neuroscience and the discoveries about how the human vision system processes visual information, computer scientists got focused on mapping those findings to build computational models emulating human object recognition [11]. Some developed models as those based on the visual primate system [18], have produced admirable results. However, it could not perform well in some situations as cluttered scenarios and/or with partially occluded objects.

So, the issues to be solved can be summarized as: object recognition at any time; and object location within the whole home. Therefore, any time a user enters his/her home, the agent is surveilling his/her possessions such that they are recognized and tracked along the time. In that way, in the case the user forgets where he/she has left a certain possession, the developed module can accurately provide its location.

In this context, the first issue to be solved refers to object recognition. Basically, it implies two different tasks: object detection and object identification. So, on the one hand, object detection refers to the ability of identifying the presence/absence of certain objects in an image. Given its importance for a wide range of visual applications, research on this topic has taken a number of forms. For instance, several approaches have built models of the objects to be detected (e.g. [9, 24]). Mainly, these techniques consist of learning a model of the interest objects in advance from a large set of training images. These saved images show the object in different poses and from different viewpoints. Although it could be a good solution for some kind of objects, it is difficult to learn models of objects with a high dimensionality and/or with a rich variability in their motion. On the contrary, other approaches have been proposed depending on the considered visual features such as shape [21], colour [19, 23], texture [4], appearance [5], motion [6], patch [12] or gradient [10]. Nevertheless, it is difficult to design a robust method based on visual features when working on real scenarios where different factors such as occlusions, changes in illumination, different viewpoint, or dynamic backgrounds (e.g. waving trees, moving curtains, or blinking screens), can make them miserably fail. As a solution, background removal algorithms have been developed (e.g. [2, 25, 16]). Basically, a statistical background model, built after observing a scene during several seconds, is used to identify objects of interest. Nevertheless, these techniques present some drawbacks to be overcome such as everything seen when the background model is being built is considered background; no sudden changes in illumination occur during the whole experiment; an object which appeared in the scene and then stopped moving for a period of time can be modelled as a part of the background; and, the resulting binary image segmentation of the image only highlights regions of non-stationary objects.

As a consequence, a mechanism for object detection in real-life scenarios has to be deployed. Keeping in mind that objects to be identified are left to any place by a person, motion is considered the primary cue for object detection. That is, this method is based on the idea that a person is leaving their possessions while they are moving around their home. Therefore, we have implemented a new approach [15, 14] able of managing a training period with foreground objects, adapting to minor dynamic, uncontrolled changes (e.g. the passage of time, blinking of screens or shadows), adapting to sudden changes in illumination and distinguishing between target objects and background in terms of motion and motionless situations.

On the other hand, once objects are detected in a captured image, object identification is the next step. On this matter, the key issue is determining which characteristic visual information allows us to uniquely identify such target objects. For that reason, although motion allows the system to robustly detect possessions in an image sequence, other features must be used. Despite the wide variety of existing approaches based on visual features [3], they are inappropriate for our purpose because they provide a variable number of features to describe an object. Thus, a high number of features results in a high space-consuming application, whilst a

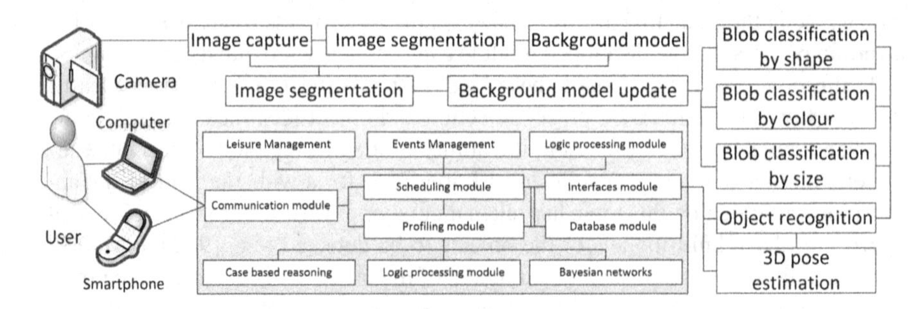

Fig. 2 iGenda and Object Recognition system overview

low one would provide inaccuracy. Moreover, the detection of substantial occlusion levels requires a large number of key points leading to a high computational cost. What is more, the relationship existing between the number of visual features and the number of objects to be tracked constrains the application. For that reason, in this work, shape, colour and size are used. Consequently, our object representation is composed of these three visual features that are modelled taking into account the variability of their values due to viewpoint changes and different illumination conditions.

The final stage of this process refers to object location within user's home. In this case, the geometry of multiple cameras is exploited to obtain the 3D object pose. Therefore, considering the most common lost objects (i.e. the keys, the wallet, the glasses, etc.), a distinctive representation of each of them is being designed and used by the visual agent. So, it can extract information about possession location from the visual input provided by a heterogeneous set of cameras composed of colour traditional and fisheye devices.

5 Conclusions and Future Work

In this paper, a modular system to aid people in remembering their scheduled tasks as well as all the objects required to perform those tasks, is presented. For that, a vision module for object recognition and location is integrated in a multi-agent home platform. This integration is based on the agent concept, which allows to overcoming the great difference between the module internal data-flow. In that way, the proposed system lays the foundations for a compliant AAL ecosystem aimed at assuring the fusion of singularly developed projects with the purpose of getting new features in different environments and platforms.

Therefore, on the one hand, the multi-agent home platform will handle the information about the requested object, searching relations in the user information database, and managing the user interfaces to accommodate new information. An aftereffect can be the introduction of a user's home map in the interface where the object in need can be easily located by showing in which part of the house it is.

On the other hand, the vision module will recognize and locate the target objects. So, while object recognition considers object shape, colour and size from a segmented image based on a background model, the object location exploites the geometry of the camera set to properly estimate 3D object poses.

The future work will be in terms of developing the communication structures and the intergration with the scheduling system. New agents and modules must be built to process the new incoming data and the bilateral trade of information. The key issue of the development is that the current core systems of both modules do not have to be changed to receive the new agents. The technological aim is to change the least possible the developed systems and to implement new features. Therefore, validating the proposal of building a stable AAL platform establishes two different projects.

Acknowledgments. This work is partially funded by National Funds through the FCT-Fundação para a Ciencia e a Tecnologia (Portuguese Foundation for Science and Technology) within projects PEst-OE/EEI/UI0752/2011, and PEst-C/CTM/LA0025/2011.

Project AAL4ALL, co-financed by the European Community Fund FEDER through COMPETE - Programa Operacional Factores de Competitividade (POFC). Authors also want to thank the support received from the Ministerio de Ciencia e Innovación (DPI2011-27846), by Generalitat Valenciana (PROMETEO/2009/052) and by Fundació Caixa Castelló-Bancaixa (P1-1B2011-54).

References

1. Acampora, G., Loia, V.: A dynamical cognitive multi-agent system for enhancing ambient intelligence scenarios. In: IEEE Int. Conf. on Fuzzy Systems, pp. 770–777 (2009)
2. Ahmed, S., El-Sayed, K., Elhabian, S.: Moving object detection in spatial domain using background removal techniques - state-of-art. Recent Patents on Computer Sciencee 1(1), 32–54 (2008)
3. Bay, H., Ess, A., Tuytelaars, T., Van Gool, L.: Speeded-up robust features (surf). CVIU 110(3), 346–359 (2008)
4. Bileschi, S.M., Wolf, L.: A unified system for object detection, texture recognition, and context analysis based on the standard model feature set. In: British Machine Vision Conference, vol. 83, pp. 1–10 (2005)
5. Borotschnig, H., Paletta, L., Prantl, M., Pinz, A.: Appearance-based active object recognition. Image and Vision Computing 18(9), 715–727 (2000)
6. Ciliberto, C., Pattacini, U., Natale, L., Nori, F., Metta, G.: Reexamining lucas-kanade method for real-time independent motion detection: Application to the icub humanoid robot. In: IEEE/RSJ Int. Conf. on Intelligent Robots and Systems, pp. 4154–4160 (2011)
7. Costa, A., Castillo, J.C., Novais, P., Fernández-Caballero, A., Simoes, R.: Sensor-driven agenda for intelligent home care of the elderly. Expert Systems with Applications 39(15), 12,192–12,204 (2012), doi:10.1016/j.eswa, 04.058
8. Costa, A., Novais, P., Corchado, J.M., Neves, J.: Increased performance and better patient attendance in an hospital with the use of smart agendas. Logic Journal of IGPL 20(4), 689–698 (2012), doi:10.1093/jigpal/jzr021
9. Cristinacce, D., Cootes, T.F.: Boosted regression active shape models. In: British Machine Vision Conference, vol. 2, pp. 880–889 (2007)

10. Dalal, N., Triggs, B.: Histograms of oriented gradients for human detection. In: CVPR, vol. 1, pp. 886–893 (2005)
11. Guido, D.: AP Neuroscience Course (2011), http://www.rhsmpsychology.com/
12. Keysers, D., Deselaers, T., Breuel, T.M.: Optimal geometric matching for patch-based object detection. ELCVIA 6(1), 44–54 (2007)
13. Marques, V., Costa, A., Novais, P.: A dynamic user profiling technique in a AmI environment. In: World Congress on Information and Communication Technologies, pp. 1247–1252. IEEE (2011), doi:10.1109/WICT.2011.6141427
14. Martinez-Martin, E., del Pobil, A.P.: Robust Motion Detection in Real-Life Scenarios, springerbr edn. Springer Briefs in Computer Science. Springer, London (2012), doi:10.1007/978-1-4471-4216-4
15. Martínez-Martín, E., del Pobil, A.P.: Robust object recognition in an unstructured environment. In: Lee, S., Cho, H., Yoon, K.-J., Lee, J. (eds.) Intelligent Autonomous Systems 12. AISC, vol. 193, pp. 705–714. Springer, Heidelberg (2012)
16. Migliore, D.A., Matteucci, M., Naccari, M.: A revaluation of frame difference in fast and robust motion detection. In: 4th ACM Int. Workshop on Video Surveillance and Sensor Networks - VSSN 2006, p. 215. ACM Press, New York (2006)
17. Radde, S., Freitag, B.: Using Bayesian Networks To Infer Product Rankings From User Needs. In: UMAP 2010 Workshop on Intelligent Techniques for Web Personalization and Recommender Systems (2010)
18. Serre, T., Wolf, L., Poggio, T.: Object recognition with features inspired by visual cortex. In: CVPR, vol. 2, pp. 994–1000 (2005)
19. Shahbaz Khan, F., Anwer, R.M., van de Weijer, J., Bagdanov, A.D., Vanrell, M., Lopez, A.M.: Color attributes for object detection. In: CVPR, pp. 3306–3313 (2012)
20. Tazari, M.R., Wichert, R., Norgall, T.: Towards a unified ambient assisted living and personal health environment. In: Wichert, R., Eberhardt, B. (eds.) Ambient Assisted Living, vol. 63, pp. 141–155. Springer, Heidelberg (2011)
21. Toshev, A., Taskar, B., Daniilidis, K.: Shape-based object detection via boundary structure segmentation. IJCV 99(2), 123–146 (2012)
22. Ughetti, M., Trucco, T., Gotta, D.: Development of agent-based, peer-to-peer mobile applications on android with jade. In: The 2nd Int. Conf. on Mobile Ubiquitous Computing, Systems, Services and Technologies, pp. 287–294 (2008)
23. Urdiales, C., Dominguez, M., de Trazegnies, C., Sandoval, F.: A new pyramid-based color image representation for visual localization. Image and Vision Computing 28(1), 78–91 (2010)
24. Urtasun, R., Fleet, D.J., Fua, P.: Temporal motion models for monocular and multiview 3d human body tracking. CVIU 104(2-3), 157–177 (2006)
25. Varcheie, P.D.Z., Sills-Lavoie, M., Bilodeau, G.A.: An efficient region-based background subtraction technique. In: Canadian Conference on Computer and Robot Vision, pp. 71–78 (2008)
26. Vardasca, R., Simoes, R.: Needs and opportunities in ambient assisted living in portugal. In: 2nd Int. Living Usability Lab Workshop on AAL Latest Solutions, Trends and Applications, AAL 2012, in Conjunction with BIOSTEC 2012, pp. 100–108 (2012)
27. Watson, I.: An introduction to case-based reasoning. In: Watson, I.D. (ed.) UK CBR 1995. LNCS, vol. 1020, pp. 1–16. Springer, Heidelberg (1995)

Efficient People Counting from Indoor Overhead Video Camera

Juan Serrano-Cuerda, José Carlos Castillo,
Marina V. Sokolova, and Antonio Fernández-Caballero

Abstract. This article introduces a system for real-time people counting. People counting is a challenging topic in the surveillance domain. The proposed system is built from INT^3-Horus, a multi-agent based framework for intelligent monitoring and activity interpretation. The system uses an indoor overhead video camera that detects people moving freely in a hall or room. The people counting system is flexible in detecting individuals as well as groups. Counting is independent of the trajectories and possible occlusions of the humans present in the scene. The initial results offered by the system are very promising in terms of specificity, sensitivity and F-score.

Keywords: People counting system, Overhead video camera setup, INT^3-Horus framework, Multi-agent system.

1 Introduction

People counting has been widely addressed during the last few years, mainly for surveillance applications [1], [11]. This paper is focused on a system that calculates in real-time the number of people that are present in a given scenario, monitored by an indoor overhead video camera overlooking a scenario such as a hall or room [13]. People move freely, as there is no clear entrance/exit at the monitored scene. Also, there is no initial limitation in the number of people to be detected; single

Juan Serrano-Cuerda · Marina V. Sokolova · Antonio Fernández-Caballero
Universidad de Castilla-La Mancha,
Departamento de Sistemas Informáticos & Instituto de Investigación
en Informática de Albacete, 02071-Albacete, Spain
e-mail: Antonio.Fdez@uclm.es

José Carlos Castillo
Instituto de Sistemas e Robótica, Instituto Superior Técnico,
1049-001 Lisbon, Portugal

J.B. Pérez et al. (Eds.): *Trends in Prac. Appl. of Agents & Multiagent Syst.*, AISC 221, pp. 129–137.
DOI: 10.1007/978-3-319-00563-8_16

humans appear in the scenario and also groups of people are allowed. An uncalibrated overhead camera is used to avoid the majority of occlusions present in lateral video camera installations.

The people counting system described in this paper has been developed from INT^3-Horus, a multi-agent based framework for intelligent monitoring and activity interpretation. The paper introduces a description of the INT^3-Horus framework, its particularization to develop the people counting system, and some promising results of the setup of the indoor overhead camera application.

2 From INT^3-Horus Framework to People Counting System

This section starts with a general description of INT^3-Horus. Then, the general structure of the people counting system is described. Finally, each level used for people counting from an indoor overhead camera is described extensively, with special focus in the algorithms developed to perform this specific task.

2.1 The INT^3-Horus Framework

INT^3-Horus is conceived as a multi-agent based framework to carry out monitoring and activity interpretation tasks [6, 7, 8]. This is an ambitious goal given the huge variety of scenarios and activities that can be faced [10, 3, 16]. The framework establishes a set of operation levels where clearly defined input/output interfaces are defined. Inside each level, a developer places his/her code, encapsulated in a module in accordance with the operation performed. Although a set of levels are proposed in INT^3-Horus to cover all the steps of a generic multi-sensor and activity interpretation system [12, 15, 5], the philosophy underlying the framework allows a flexible set of levels to be adapted to a given final system [2, 4].

The framework infrastructure as well as the modules layout is based on the Model-View-Controller (MVC) paradigm [14], which allows to isolate the user interface from the logical domain for an independent development, testing and management. The MVC paradigm divides an application into three main entities, defining their main roles as well as the connections among them. *Model* manages the application data, initializes objects and provides information about the application status. In event-driven systems, the *Model* informs the *View* and the *Controller* about information changes. *View* provides a representation of the *Model* information (performing just simple operations) to fit user requirements. Finally, *Controller* receives the inputs to the application and interacts with the *Model* to update its objects, and with the *View* to represent the new information.

Despite of the many benefits provided by the MVC paradigm, the union of the business logic and the data model presents a drawback when it comes to add new functionalities to the framework. To solve this issue the traditional MVC is extended for INT^3-Horus, creating a new component to house each module's specific operation (see Fig. 1 on the right). This way, each framework user receives a template with a series of components which are already integrated into INT^3-Horus. The main task

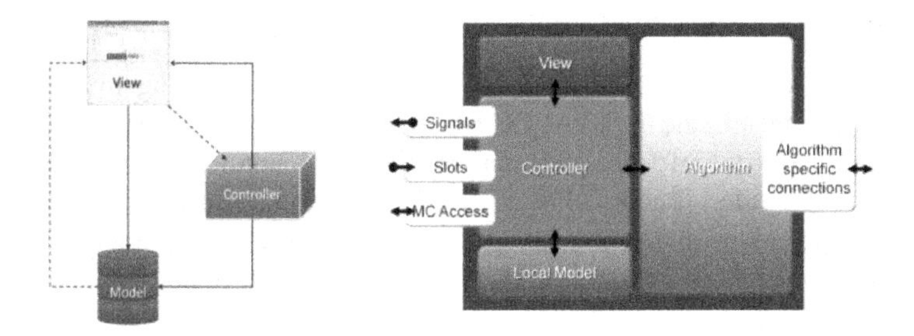

Fig. 1 Model-View-Controller representation. The left image shows the traditional MVC layout. The right image shows the extended MCV.

for the integration is to introduce the code into the component named "Algorithm" and tune up the rest of components if needed (e.g. adding controls to the module's *View* or data structures to the local *Model*). The module's controller provides the connections to the framework and the access to the global data model as well as the signals to control the execution.

In this sense, the framework allows easy code integration, providing users with module templates to put their code into them. These templates already have the necessary connections to access the rest of INT^3-Horus components, not only the data model or the user interface, but also the controller to trigger each module's execution.

Together with the easy addition of new functionalities, a state of the art framework for monitoring and activity detection must take into account several information sources (sensors) and INT^3-Horus is not an exception. These sources are mainly related to image sensors since they are the most widespread for monitoring tasks; but other sensor technologies, like commercial sensors and wireless sensor networks (WSNs), are also integrated to show the generic purpose of INT^3-Horus.

2.2 INT^3-Horus Levels for People Counting

The main goal of an efficient people counting system is to obtain the number of humans inside the field of view of a camera. In this particular case, we describe a system based on an indoor overhead camera with the highest possible accuracy when counting people. For this purpose, four processing levels are selected from INT^3-Horus framework. The lower one is in charge of collecting data from a overhead camera (*acquisition*). The next level, *segmentation*, uses an approach based on background subtraction to isolate the humans in the scene. Some filtering and heuristics are applied in the next level, *blob detection*, to enhance segmented humans, dealing with false positives and splitting groups of people into individuals.

2.2.1 The Data Acquisition Level

At *acquisition* level, a specific module is in charge of capturing images from Axis cameras. This module uses VAPIX (`http://www.axis.com/techsup/cam_servers/dev/index.htm`), an HTTP-based application programming interface that provides functionality for requesting images, controlling network camera functions (pan-tilt-zoom, relays, etc.) and setting/retrieving internal parameter values. The proposed people counting system only needs images obtained from a networked camera.

2.2.2 The Segmentation Level

The main objective of the *segmentation* level is to perform the initial detection of the humans present in the scene. An adaptive Gaussian background subtraction is performed on input image I_Z obtained from the indoor overhead camera, as shown in Fig. 2(a). The subtraction is based on the OpenCV implementation of a well-known algorithm [9]. The algorithm builds an adaptive model of the scene background based on the probabilities of a pixel to have a given color level. An example of this model is shown in Fig. 2(b). A shadow detection algorithm, based on the computational color space used in the background model, is also used. After the background segmentation is performed, an initial background segmentation image I_B is obtained as shown in Fig. 2(c).

However, the resulting image contains some noise which must be eliminated. For this, an initial threshold θ_0 (experimentally fixed as a 16th of the number of possible gray levels in the image) is applied, as shown in equation (1):

$$I_{Th}(x,y) = \{ \text{ min , if } I_S(x,y) \leq \theta_0, \text{ max , otherwise} \tag{1}$$

and

$$\theta_0 = \frac{\text{max}}{16} \tag{2}$$

where *min* is fixed to 0 (since we are obtaining binary images) and *max* is the maximum gray level value that a pixel can have in I_B (e.g. 255 for an 8-bit image).

After this operation, two morphological operations, namely opening and closing, are performed to eliminate the remaining noise of the image, obtaining I_S as shown in Fig. 2(d). After the first noise reduction, the number of white pixels (corresponding to possible humans) is counted in the image. If this value is greater than a 50% of the area of the image during a predefined amount of time Δt (usually one second), it is estimated that a big lighting change has occurred in the scene (e.g. a light switch turned on/off or a door was opened/closed). In this case, the algorithm is initialized to build a new background based on the new lighting conditions of the scene.

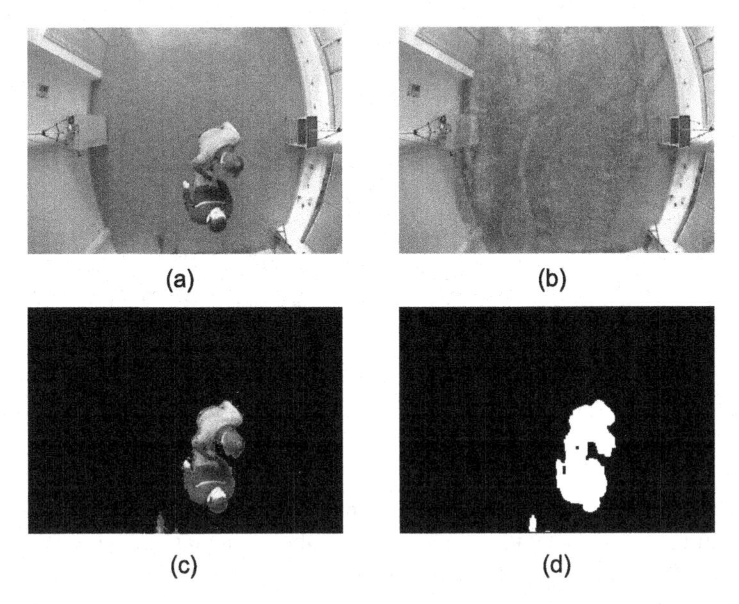

Fig. 2 Images generated by the segmentation level. (a) Current frame. (b) Background model. (c) Foreground. (d) Segmented image.

2.2.3 The Blob Detection Level

Now, human candidates must be extracted from binary image I_S, paying special attention to the existence of groups of people. For this purpose, the concept of region of interest (ROI) is explained. A ROI is defined as the minimum rectangle containing a human. It can be characterized by a pair of coordinates (x_{min}, y_{min}),(x_{max}, y_{max}), corresponding to the upper-left and lower-right limits of the ROI, respectively. All detected ROIs are used to annotate the humans detected in the scene in a list L_B.

In first place, human candidates are extracted from the scene. With this objective, connected components (blobs) are extracted from I_S. Next, blobs with a ROI area lower than A_{min} (with a value experimentally fixed according to scene features such as the scene area or the height where the camera is placed) are discarded. A new area threshold A_G is also established based on similar factors. Blobs with a ROI area lower than A_G are considered to contain a single human and the ROI containing it is enlisted in L_B. Otherwise, blobs B_G with a ROI area greater than A_G are analyzed separately, since they are considered to possibly contain a group of humans. Now, each human belonging to these groups is extracted individually. To do so, a new subimage I_G is created containing the ROI delimiting B_G, as shown in Fig. 3(a). Then, a new series of morphological openings are performed, since occlusions are less frequent in an overhead view than in a lateral view, obtaining a new image I_G. An example of the result of these operations is offered in Fig. 3(b). Next, blobs are searched in this new image. Now, blobs with an area greater than A_{min} are annotated in a list of group blobs L_{B_G}, whilst the others are discarded. If, at the end of the search, L_{B_G} is empty, the original ROI with the blob B_G is enlisted as a single human;

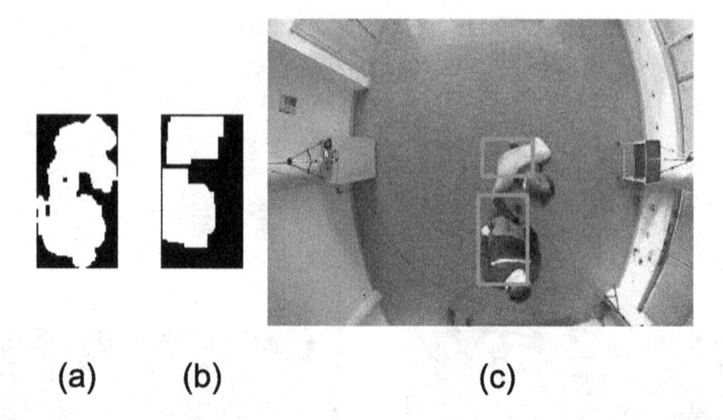

Fig. 3 Results of the blob detection level. (a) Original ROI. (b) Separated ROIs. (c) Final Result.

but, it will be marked as a possible group that could not be separated. Finally, the blobs from L_{B_G} are enlisted in L_B, where the number of humans in the scene (people counting) is the number of blobs contained in L_B. The detected humans in the scene are shown in Fig. 3(c) for this running example.

3 Data and Results

Three different video sequences were recorded from an Axis camera to test our proposal. The first sequence shows different people walking along a hallway (individually or in groups of two or three individuals). Generally, the people do not stop and do not cross their paths, except for the final frames of the video, where two people meet and talk for a while in the center of the scene and another person approaches them (see Fig. 4(a)). The second sequence is similar to the first one, although more occlusions appear as different people intersect their paths. Another meeting takes place in this second video; this time, there are three people remaining still in the scene for a minute without being added to the background model. An example of the group separation in this sequence is shown in Fig. 4(b). The final sequence is the most complex one. In this video, up to five people appear in the scene, crossing and intersecting their paths, which results in a greater amount of occlusions than in the previous sequences. They also meet for two minutes without being added to the background, and partially occluding themselves. Nevertheless, they are detected most of the time as shown in Fig. 4(c).

Table 1 shows some quantitative results extracted from the three sequences. In order to evaluate the performance, we have used measures of specificity, sensitivity and F-score. These are usual statistics in image processing, calculated as shown in equations (3), (4) and (5) respectively.

$$specificity = \frac{TP}{TP+FP} \tag{3}$$

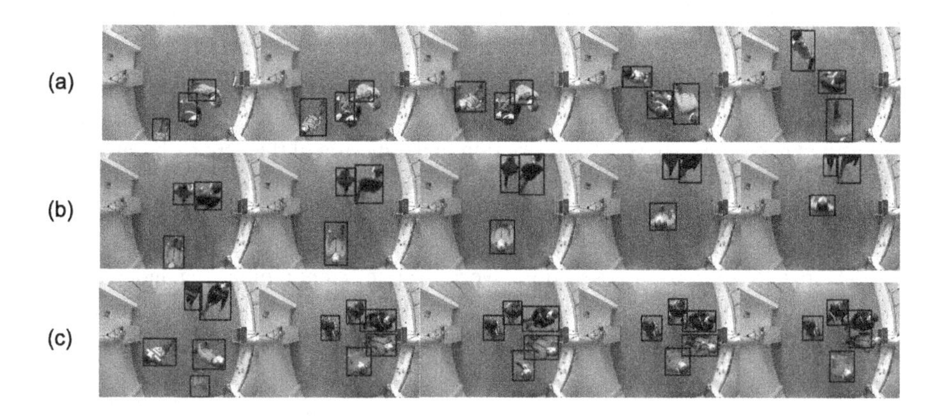

Fig. 4 Qualitative results for the three recorded sequences. (a) Video 1. (b) Video 2. (c) Video 3.

Table 1 Results of people counting in different video sequences

Sequence	Humans in the sequence	Humans detected	TP	FP	FN	Specificity	Sensitivity	F-score
1	1060	1024	1001	23	59	0,944	0,978	0,960
2	1698	1702	1656	46	42	0,975	0,973	0,974
3	3199	3274	3128	146	71	0,978	0,955	0,966
Total	5957	6000	5785	215	172	0,971	0,964	0,967

$$sensitivity = \frac{TP}{TP + FN} \tag{4}$$

$$F - score = \frac{2 * specificity * sensitivity}{specificity + sensitivity} \tag{5}$$

where TP (true positives) is the number of correct detections in the scene, FP (false positives) is the number of humans detected but actually not present, and FN (false negatives) is the number of humans present in the scene that have not been detected by our algorithm.

Notice that the results are really outstanding. The first sequence shows worse results since several small lighting changes took place during the recording. Also small misdetections have greater impact in the final statistics as less humans appear in this sequence than in the later ones. It is also important to highlight that the results of the final sequence show the difficulty of the video, since a lot of humans appear and they are occluding themselves most of the time.

4 Conclusions

This paper has introduced an efficient people counting system. The system based on an indoor overhead camera counts the number of people that are present in a given scenario in real-time. There is no restriction in the motion of the people. Even, there is no limitation in the number of people to be detected. The people counting system accepts individual as well as groups of people.

The people counting system described in this paper has been developed from INT^3-Horus, a multi-agent based framework for intelligent monitoring and activity interpretation. The paper has demonstrated the usefulness of the framework and the accuracy of the developed system.

Acknowledgements. This work was partially supported by Spanish Ministerio de Economía y Competitividad / FEDER under TIN2010-20845-C03-01 grant. This work was also partially sponsored by the project CMU-PT/SIA/0023/2009 under the Carnegie Mellon Portugal Program and its Information and Communications Technologies Institute.

References

1. Boltes, M., Seyfried, A.: Collecting pedestrian trajectories. Neurocomputing 100, 127–133 (2013)
2. Carneiro, D., Castillo, J.C., Novais, P., Fernández-Caballero, A.: Multimodal behavioural analysis for non-invasive stress detection. Expert Systems with Applications 39(18), 13376–13389 (2012)
3. Castillo, J.C., Fernández-Caballero, A., Serrano-Cuerda, J., Sokolova, M.V.: Intelligent monitoring and activity interpretation framework - INT3-Horus ontological model. In: Advances in Knowledge-Based and Intelligent Information and Engineering Systems, pp. 980–989 (2012)
4. Costa, A., Castillo, J.C., Novais, P., Fernández-Caballero, A., Simoes, R.: Sensor-driven agenda for intelligent home care of the elderly. Expert Systems with Applications 39(15), 12192–12204 (2012)
5. Fernández-Caballero, A., Castillo, J.C., Rodríguez-Sánchez, J.M.: Human activity monitoring by local and global finite state machines. Expert Systems with Applications 39(8), 6982–6993 (2012)
6. Gascueña, J.M., Fernández-Caballero, A.: On the use of agent technology in intelligent, multisensory and distributed surveillance. The Knowledge Engineering Review 26(2), 191–208 (2011)
7. Gascueña, J.M., Fernández-Caballero, A., López, M.T., Delgado, A.E.: Knowledge modeling through computational agents: application to surveillance systems. Expert Systems 28(4), 306–323 (2011)
8. Gascueña, J.M., Navarro, E., Fernández-Caballero, A.: Model-driven engineering techniques for the development of multi-agent systems. Engineering Applications of Artificial Intelligence 25(1), 159–173 (2012)
9. KaewTraKulPong, P., Bowden, R.: An improved adaptive background mixture model for real-time tracking with shadow detection. In: Video Based Surveillance Systems: Computer Vision and Distributed Processing, pp. 1–5 (2001)
10. Kieran, D., Yan, W.: A framework for an event driven video surveillance system. Journal of Multimedia 6(1), 3–13 (2011)

11. Moreno-Garcia, J., Rodriguez-Benitez, L., Fernández-Caballero, A., López, M.T.: Video sequence motion tracking by fuzzification techniques. Applied Soft Computing 10(1), 318–331 (2010)
12. Pavón, J., Gómez-Sanz, J., Fernández-Caballero, A., Valencia-Jiménez, J.J.: Development of intelligent multi-sensor surveillance systems with agents. Robotics and Autonomous Systems 55(12), 892–903 (2007)
13. Rao, R., Taylor, C., Kumar, V.: Experiments in robot control from uncalibrated overhead imagery. In: Ang, M.H., Khatib, O. (eds.) Experimental Robotics IX. STAR, vol. 21, pp. 491–500. Springer, Heidelberg (2006)
14. Reenskaug, T.: Thing-model-view-editor an example from a planning system. XEROX PARC Technical Note (May 1979)
15. Rivas-Casado, A., Martinez-Tomás, R., Fernández-Caballero, A.: Multiagent system for knowledge-based event recognition and composition. Expert Systems 28(5), 488–501 (2012)
16. Sokolova, M.V., Castillo, J.C., Fernández-Caballero, A., Serrano-Cuerda, J.: Intelligent monitoring and activity interpretation framework - INT3-Horus general description. In: Advances in Knowledge-Based and Intelligent Information and Engineering Systems, pp. 970–979 (2012)

Modeling Intelligent Agents to Integrate a Patient Monitoring System

Gabriel Pontes, Carlos Filipe Portela, Rui Rodrigues,
Manuel Filipe Santos, José Neves, António Abelha, and José Machado*

Abstract. ICU units are a good environment for the application of intelligent systems in the healthcare arena, due to its critical environment that require diagnose, monitor and treatment of patients with serious illnesses. An intelligent decision support system - *INTCare*, was developed and tested in CHP, a hospital in Oporto, Portugal. The need to detect the presence or absence of the patient in bed, in order to stop the collection of redundant data concerning about the patient vital status led to the development of an RFID locating and monitoring system - PaLMS, able to uniquely and unambiguously identify a patient and perceive its presence in bed in an ubiquitous manner, making the process of data collection and alert event more accurate. An intelligent multi-agent system for integration of PaLMS in the hospital's platform for interoperability (AIDA) was also developed, using the characteristics of intelligent agents for the communication process between the RFID equipment, the INTCare module and the Patient Management System, using the HL7 standard embedded in agent behaviours.

Keywords: Ambient Intelligence, Intelligent Agents, Multi-Agent Systems, RFID, Patient Monitoring, Intensive Care Units, Medical Informatics.

Gabriel Pontes · Rui Rodrigues · José Neves ·
António Abelha · José Machado
University of Minho,
Computer Science and Technology Centre (CCTC),
Campus de Gualtar,
Braga, Portugal
e-mail: `jmac@di.uminho.pt`

Carlos Filipe Portela · Manuel Filipe Santos
University of Minho, ALGORITMI Center,
Campus de Azurem, Guimarães, Portugal

* Corresponding author.

J.B. Pérez et al. (Eds.): *Trends in Prac. Appl. of Agents & Multiagent Syst.*, AISC 221, pp. 139–146.
DOI: 10.1007/978-3-319-00563-8_17

1 Introduction

HealthCare is an area of constant development towards better patient treatments and better quality of service provided. New demands are constantly arising, and computer technology is in the frontline of the responses to those demands, not only in the field of medical assistance, but also in an administrative and organizational point of view. Yet, classical computation paradigms fall short when trying to model an environment with a large number of users and complex processes and interactions [4]. *Multi-Agent Systems* (MAS) stands as an emerging technology focussing on the modelling, design and development of complex systems. The use of intelligent agents in Medicine has been shown as a complementary technique to improve the performance of computer-based systems, in terms of interoperability, scalability and reconfigurability. Moreover, it is one of the main topics of international conferences in *Artificial Intelligence* (AI). *Centro Hospitalar do Porto* (CHP), a hospital facility in Oporto, in the north of Portugal, has developed and implemented *INTCare* [7] an intelligent decision support system (IDSS) for the *Intensive Care Unit* (ICU), aiming the real-time monitoring of patients, predicting the dysfunction or failure of six organic systems within a short period of time, and the patient outcome in order to help doctors deciding on the better treatments or procedures for the patient. This system fails when the patient is out of bed. The data generated by the monitoring system becomes redundant when there is no one to monitor, and computational resources can be spared. INTCare uses the agent paradigm in the process of collection, analysis and process of data acquired in each bed from the unit.

This article is divided into five sections. Besides this introduction section where a brief description is given about the problem being dealt with and the project description, in section 2 we talk about the technologies that stand on the basis of the project, presenting some literature background and state-of-the-art of the interoperability and HL7 as a standard to overcome that issue, and the characteristics of intelligent agents and multi-agent systems. Section 3 relates to the INTCare project developed in CHP, its main characteristics and goals. PaLMS system is described in Section 4, where the method of message exchange between agents driven by different events, triggered according to specific situations is explained and detailed. Section 5 presents some conclusions about the project and the future work expected to be done in order to improve the project itself, and some hospital developments achievable using RFID technology and HL7 messages embedded in intelligent agents.

2 Background

Even with the ongoing increase in hospitals use of computerized tools such as powerful hospital information systems and connected laboratory results, these tools are not sufficient and new technologies should support a new way of envisaging the future hospital [2].

2.1 Multi-Agent Systems

Agents and the *Agent-Oriented Programming* are concepts correlated to the field of Artificial intelligence, and their importance is growing in the health-care environment, namely in the "quest" for interoperability. Shoham, Y. described in 1993 [10] an agent as *"an entity whose state is viewed as consisting of mental components such as beliefs, capabilities, choices and commitments"*. The use of intelligent agents in Medicine has been shown as a complementary technique to improve the performance of a computorised system in terms of interoperability, scalability and reconfigurability [5].

2.2 Interoperability

One of the medical informatics' concerns nowadays is to ensure interoperability. With the grow of medical organizations, the need to manage information between many components also increases, and the most common scenario is to find such organizations with specialist sub-domains, each with its own vocabulary, knowledge base and software applications. These sub-domains contain multi-platform, multi-vendor application wrappers built around multi-variate data sources further adds to the complexity [6]. Besides, citizens move nowadays more often from one country to another for work or leisure fact that made the semantic interoperability[1] of electronic healthcare records (EHR) systems a major challenge in eHealth. The semantic interoperability was, in fact, the target of recommendations from the European Commission [3].

2.3 AIDA

AIDA stands for Agency for Integration, Diffusion and Archive of medical information, and was created by researchers both from University of Minho and CHP. AIDA can be described as an agency that provides intelligent electronic workers (agents) that present a pro-active behaviour, and are in charge of tasks such as: communications among the sub-systems that make the whole one, sending and receiving information (e.g. medical or clinical reports, images, collections of data, prescriptions), managing and saving the information and answering to information requests, in time [1].

[1] Semantic interoperability of clinical information can be defined as the ability of information systems to exchange and understand clinical information independently of the system in which it was created.

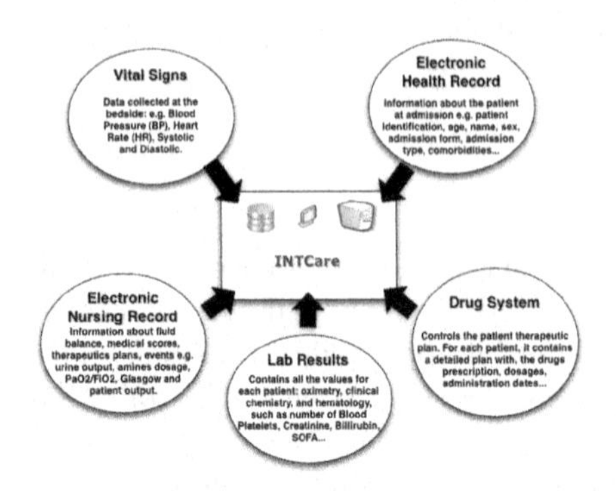

Fig. 1 Illustration of the five different types of input data to the INTCare system

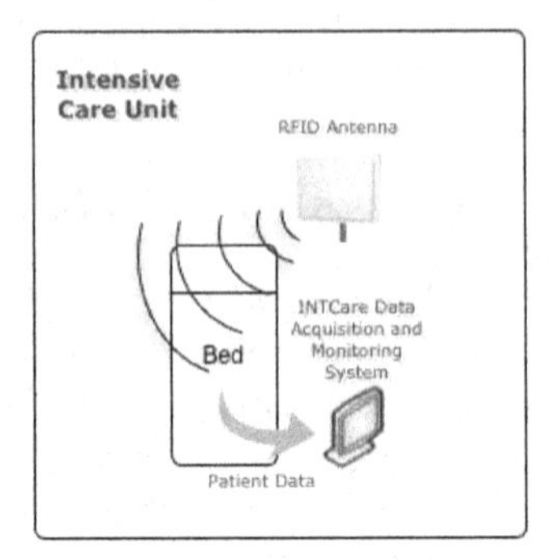

Fig. 2 ICU equipped with RFID antenna and screens to data consult and analysis from the medical staff

3 INTCare

INTCare is an Intelligent Decision Support System (IDSS) developed for Intensive Medicine, with the main goal of predicting the organ failure and outcome in real time, improving the health care by allowing the physicians to take a pro-active attitude in the best interest of patients. INTCare, developed and fully implemented in CHP, is capable of predicting organ failure probability, the outcome of the patient for the next day, as well as the best suited treatment to apply. Due to the new fine-grained time response requirements, it is very useful to have models to predict values for the next hour, which means that the system should be adapted to real-time data [9] [7].

INTCare project has five data sources concerning about the patient monitoring process and data management, described in Figure 1 [7] [11].

This system uses a Knowledge Discovery in Databases (KDD) process. In order to model information for KDD processing, there are some requirements that should be met: [9] [7]: Online Learning, Real-time, Adaptability, Data Mining Models, Decision Models, Optimization and Intelligent Agents.

4 PaLMS

When patient leaves the ICU for an exam, for the operating room, or any other place while technically he hasn't discharged the unit, a problem occurs. The alert system developed inside the INTCare to alert the medical professionals about the patient condition enters in a warning state, but no patient is in the bed. One of the causes is that some patient vital data collected gets out of the normal parameters when analyzed, thus starting the alert process. Other problem related to the patient's absence concerns about the computational resources used to analyze data in the case described earlier, when the patient is not in the unit. Analysis of vital signs are made, information is stored, but this data is simply redundant.

A system that perceives the patient presence in bed is needed. This perception of the physical environment by computation entities invokes Ambient Intelligence. We propose an intelligent system embedded in the environment, able to detect the patient presence in bed through a wireless technology: the radio-frequency identification. At this first stage of implementation, a single antenna is installed on the bedside of each bed in the ICU, assuring the full-bed coverage by the RFID beam emitted by the antenna. A study may be realized in order to test the RFID accuracy using a single antenna, and the possibility of placing two antennas to optimize the process is also to be considered, depending on the study's results. In both cases, a patient in the bed can be monitored, by simply using a bracelet containing one RFID tag. This way, the patient can be identified uniquely and unambiguously.

This method of patient identification and his detection in the unit must have a way to be connected to the hospital's Patient Management System (PMS)[2], so that the process of patient admission and discharge, the RFID reading tag process, the INTCare data acquisition and the alert system all work together, in a synchronous way. When a patient is in the bed, the whole system must: 1) gather the patient vital sign data in order to be monitored and processed; 2) have its alert system enabled to detect abnormalities in the patient's vital signs; 3) have the RFID monitoring the patients presence in the bed, and if not, stop the earlier processes.

4.1 Event-Based Model

To manage all the communication between the different systems, we have chosen a multi-agent system, using intelligent agents and their unique capabilities to handle the information exchange. A system of messages are in the basis of the communications, informing each system about possible changes about the patient. These changes were considered in the present work as events. The five far most usual events occurring in the patient cycle inside the healthcare facility, since the patient arrival until his departure, are considered to be admission, discharge, transfer, leave of absence and return from leave of absence.

4.2 Multi-Agent System

The process of communication chosen, as it was said before, was a message exchange between entities, receiving and reacting accordingly to the characteristic of the message. This was the main reason under the option for intelligent agents. Their behaviour capabilities enable them to have different

[2] In Portugal, the patient management system adopted and implemented in almost all of healthcare institutions is SONHO [8]

Table 1 Description of the events used in PaLMS

Event	Description
Admission Event (AE)	Event sent by AIDA to the ICU informing the admission of a patient, and an unused EPC code to be inserted in the RFID tag, to place in the patients' arm.
Discharge Event (DE)	Event sent by AIDA to the ICU informing the discharge of the patient, and the corresponding RFID's EPC code.
Transfer Event (TE)	Event sent by AIDA to the ICU informing the transfer of a patient to another unit, and the corresponding RFID's EPC code.
Leave of Absence Event (LoAE)	Event sent by AIDA to the ICU informing the leave of absence of the patient, as he goes, for instance, to take an exam outside the ICU, and the corresponding RFID's EPC code.
Return from Leave of Absence Event (RLoAE)	Event sent by AIDA to the ICU informing the return of the patient to the ICU, and the corresponding RFID's EPC code.
Warning Event (WE)	Event sent by the RFID equipment to the ICU warning about the fail to detect the tag in the ICU bed, and the corresponding RFID's EPC tag.

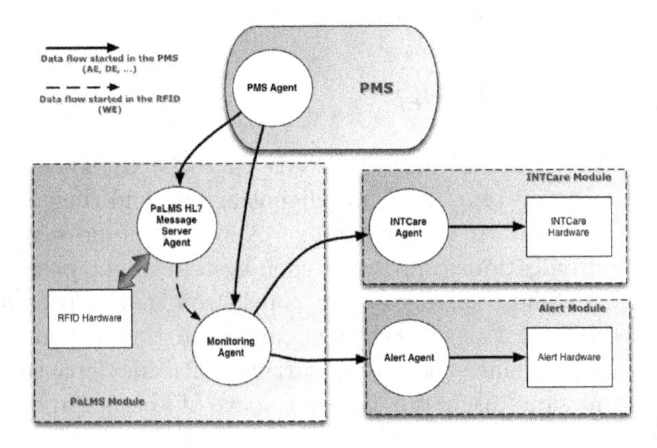

Fig. 3 Description of the communication of the three modules inside AIDA

reactions according to the need expressed to them. In our case, when we send a message informing the admission of a patient, the entity must react in a different manner than if we are dealing with a discharge event.

5 Conclusions and Future Work

Being still in a development stage, PaLMS brings together many emergent technologies in healthcare in order to solve the problem we were dealing with: the need to recognize and monitor the patient presence in a specific bed, optimizing another system for medical assistance. Further to the development stage, tests will be done to evaluate the potenciality of such type of monitoring system and the possible failures. Although, multi agent systems bring many advantages if properly adapted, assuring the system persistence, versatility and reliability. Having HL7 embedded in the process of message exchange has so far bring no disadvantages nor problems in the process.

This type of monitoring system using RFID can be applied and tested in other units, to monitor the patients movements inside the hospital or in any particular unit, to track a specific doctor needed at that time, etc. As the main framework is developed, such RFID systems are able to be inserted and implemented in the hospital facility, speeding up the process. The fact that intelligent agents are evolved with the hospital core of information, such as the patient management system, enables the hospital to get this kind of information collected by RFID systems available and capable of communication with any other system or unit, making the whole process interoperable.

An aspect that will require attention is the data persistence between the agent communications. Since we are dealing with patients, specially in the ICU, having their life at risk, such system cannot be unprepared to possible failures, systems containing the agents being shut down, losing the information about the start or stop or readings of the RFID or INTCare. Therefore, a way to store information about the agent status, agents received and sent messages, etc, is being prepared through data storage in databases.

Acknowledgements. This work is financed with the support of the Portuguese Foundation for Science and Technology (FCT), with the grant SFRH/BD/70549/2010 and within project PEst-OE/EEI/UI0752/2011.

References

1. Abelha, A., Analide, C., Machado, J., Neves, J., Santos, M., Novais, P.: Ambient intelligence and simulation in health care virtual scenarios. In: Camarinha-Matos, L., Afsarmanesh, H., Novais, P., Analide, C. (eds.) Establishing The Foundation of Collaborative Networks. IFIP AICT, vol. 243, pp. 461–468. Springer, Boston (2007)
2. Bricon-Souf, N., Newman, C.R.: Context awareness in health care: A review. International Journal of Medical Informatics 76(1), 2–12 (2007), `http://www.sciencedirect.com/science/article/pii/S1386505606000098`, doi:10.1016/j.ijmedinf.2006.01.003

3. Costa, C.M., Menárguez-Tortosa, M., Fernández-Breis, J.T.: Clinical data interoperability based on archetype transformation. Journal of Biomedical Informatics 44(5), 869–880 (2011), `http://www.sciencedirect.com/science/article/pii/S1532046411000979`, doi:10.1016/j.jbi.2011.05.006
4. Fox, J., Beveridge, M., Glasspool, D.: Understanding intelligent agents: Analysis and synthesis. AI Communications 16, 139–152 (2003)
5. Isern, D., Sánchez, D., Moreno, A.: Agents applied in health care: A review. International Journal of Medical Informatics 79(3), 145–166 (2010), doi:10.1016/j.ijmedinf.2010.01.003
6. Orgun, B., Vu, J.: Hl7 ontology and mobile agents for interoperability in heterogeneous medical information systems. Computers in Biology and Medicine 36(7-8), 817–836 (2006) Special Issue on Medical Ontologies, `http://www.sciencedirect.com/science/article/pii/S0010482505000806`, doi:10.1016/j.compbiomed.2005.04.010
7. Portela, F., Gago, P., Santos, M.F., Silva, Á.M., Rua, F., Machado, J., Abelha, A., Neves, J.: Knowledge discovery for pervasive and real-time intelligent decision support in intensive care medicine. In: Filipe, J., Liu, K. (eds.) KMIS, pp. 241–249. SciTePress (2011)
8. Ribeiro, J.V., Geirinhas, P.M.: Icnp in sonho. In: Mortensen, R.A. (ed.) ICNP in Europe: Telenurse, Technology and Informatics, pp. 131–136. IOS Press (1997)
9. Santos, M.F., Portela, F., Vilas-Boas, M., Machado, J., Abelha, A., Neves, J., Silva, A., Rua, F.: Information architecture for intelligent decision support in intensive medicine. W. Trans. on Comp. 8(5), 810–819 (2009), `http://dl.acm.org/citation.cfm?id=1558772.1558779`
10. Shoham, Y.: Agent-oriented programming. Artificial Intelligence 60(1), 51–92 (1993), `http://www.sciencedirect.com/science/article/pii/0004370293900349`, doi:10.1016/0004-3702(93)90034-9
11. Vilas-Boas, M., Gago, P., Portela, F., Rua, F., Silva, Á., Santos, M.F.: Distributed and real time data mining in the intensive care unit. In: 19th European Conference on Artificial Intelligence - ECAI Lisbon, pp. 51–55 (2010)

Towards Self-Explaining Agents

Johannes Fähndrich*, Sebastian Ahrndt, and Sahin Albayrak

Abstract. We advocate Self-Explanation as the foundation for the Self-* properties. Arguing that for system component to have such properties the underlining foundation is a awareness of them selfs and their environment. In the research area of adaptive software, self-* properties have shifted into focus pushing ever more design decisions to a applications runtime. Thus fostering new paradigms for system development like intelligent agents. This work surveys the state of the art methods of self-explanation in software systems and distills a definition of self-explanation.

Keywords: Self-Explanation, Self-*, Intelligent Agents, Self-CHOP.

1 Introduction

The development of distributed systems in heterogeneous environments is a challenging task for humans [12]. As a matter of fact, the management of such systems where different parties at different times make use of different technologies to reach their goals becomes ever more difficult. Additionally, systems can dynamically change due to the presence or absence of agents, services and/or devices leading to configuration problems as well as to the occurrence of emergent behavior meaning behavior which are not pre-programmed into the systems. To address the arising issues, developers attempt to shift evermore details to the application's runtime enabling the system to adjust their internal states as a result to exogenous and/or endogenous influences [14, 13]. In this process, the exogenous influences can be identified as the context the system is embedded in, whereas the endogenous influences stem from the system itself. Here, the identification and reaction as response to an influence depends on several self-* properties [23], where the

Johannes Fähndrich · Sebastian Ahrndt · Sahin Albayrak
DAI-Labor, Technische Universität Berlin, Ernst-Reuter-Platz 7, 10587 Berlin, Germany
e-mail: johannes.faehndrich@dai-labor.de

* Corresponding author.

J.B. Pérez et al. (Eds.): *Trends in Prac. Appl. of Agents & Multiagent Syst.*, AISC 221, pp. 147–154.
DOI: 10.1007/978-3-319-00563-8_18

initial set is known as self-CHOP (configuring, healing, optimizing, protecting) [11]. Admittedly these properties are rather high-level and can be distinguished into several basic properties, where one of this is Self-Explanation. Self-Explanation is inspired not only from biological systems but also by the field of social science. In this context, self-explanation is defined as an ability "of explaining to oneself in an attempt to make sense of new information, either presented in a text or in some other medium" [4]. Commonly, explaining events, intentions and ideas is a well-known way of communicating information in everyday life. On the one hand, the explaining entity is able to impart knowledge to some audience. On the other hand, the audience is able to understand and comprehend the explainer's intentions and they may even understand the explainer's course of actions. The goal of this work is to foster the understanding about the self-explanation property, specialized on multi agent systems where the description an agent can provide about itself, is interpreted as a explanation. Therefore we will provide an overview about the research field and the requirements we identified. In addition, we will introduce a formal definition of self-explanation and a metric enabling to decide which description is more self-explanatory.

2 Self-Explanation in a Nutshell

In the Cambridge Dictionary[1] the term to explain is defined as "to make something clear or easy to understand by describing or giving information about it". By examining this definition we notice that explaining is the act of giving information about an subject of interest to an audience with the intend to foster both the knowing and the understanding of the subject of interest. Going back to the initial set of self-* properties one can imagine that self-explanation injects momentum not only to the self-configuration but also to the other properties. Indeed, these properties can not be considered independently. Consequently, the term self-explanation has different meanings, too. Taking into account the different parties involved – agents (the system itself), developers and (end)users – we can distinguish between two sides of self-explanation. To start with, the *system side* aiming to integrate new agents autonomously into the existing infrastructure [22, 15]. Following the idea of self-explanation this means that new agents as well as existing ones are able to learn the capabilities of each other and to comprehend in which way they are able to interact (e.g. which data format and expressions match). One can imagine this process in the way a new human introduces itself into a prior unknown group of other humans by explaining its name and capabilities. Further we refer to the *human side* aiming to integrate the user into the system. As those systems are typically goal-driven, humans should be enabled to set the pursuit goals, to restrict the systems using constraints and to observe the results of the self-organization process [21, 22, 3].

However, several definitions of explanations have been proposed. Each one specialized for the needs of some domain. We will look at some of them to see how they can help defining the term. To start with, in statistics we can identify evidence

[1] Cambridge Dictionary Online, visit `http://dictionary.cambridge.org/`

weights in a Bayesian believe network as explanations [10]. These weights represent the logarithmic likelihood ratio of the influence of an observation on a specific variable. Therefore they can and indeed are used to explain in which way the occurrence of an event influences the current systems state [6]. To ease the access of humans to these statistical explanations different classes of techniques can be applied (e.g. verbal explanations [8] and graphical explanations [5]). In addition, *Druzdzel* [6] identified two categories in which such explanations can be separated: *Explanation of Assumptions* focusing on the communication of the domain model of the system and *Explanation of Reasoning* focusing on how conclusions are made from those assumptions. It might be worthwhile to transfer these categories to self-explanation since the meaning of concepts used might differ depending the exogenous or endogenous origin of the fact explained. Therefore the reasoner has to distinguish between the explanation of the system itself and how it can be interpreted related to the current context. This work focuses on the explanation of assumptions, since the audience of such an description is seen as an external system component. As those approaches are quite fundamental and thus general we further want to list more practical approaches in the agent community:

- *Braubach et al.* [2] uses the beliefs, desires and intents to formulate goals, knowledge and capabilities for a multi-agent system
- *Grüninger et al.* [9] uses First-order Logic Ontology for Web Services (FLOWS) to describe the functionalities of a service
- *Sycara et al.* [25] formulates agent and service capabilities utilizing the Input, Output, Precondition and Effect (IOPE) approach
- *Martin et al.* [19] uses the Ontology Web Language to structure the description of services

Those approaches all explain something about the subject of interest in specific domains but all lack the ability to measure the amount of information transfered by such an explanation, making it impossible to distinguish the quality of such explanations. In this work, we want to subsume those approaches in an theoretical framework building the foundation for a measure of the amount of self-explanation. For the reminder of this work we will utilize the following definition for the term self-explanation:

Definition 1. *Self-explanation identifies the capability of systems and system components to describe themselves and their functionalities to other systems, components or human beings.*

3 Towards Self-Explanatory Descriptions

In order to enable a system to be self-explaining the system has to provide information about its capabilities, interaction ways and current state. Nowadays this information are provided by e.g. service descriptions. The problem at hand here is the semi-optimal performance of AI algorithms using currently available descriptions like service matcher and planner [16]. As there are multiple improvement points (for example the reasoner, the knowledge-base, the used languages and the formalisms), self-explanatory descriptions try to improve the description side. To extend the current available description to self-explanatory description, we distinguish between three different types of information: *Syntax*, concerning the interpretation of signals, *Semantics* concerning the meaning and relationship between entities and *Pragmatics* concerning the interpretation of statements [20]. *Sooriamurthi* and *Leake* [24] follow this fragmentation and present in an early work their view point on explanations in the Artificial Intelligence (AI) research domain. The authors emphasize that the context should be incorporated in the interpretation and creation of explanations to enable systems to adapt to dynamic situations and therefore introduce the use of pragmatics as context-dependent interpretation of meanings. This is important since the explaining system might have to cope with partial observable situations while creating an explanation. In such situations the proposed approach suggest to take former explanations to guide the search for information to create a new explanation. *Leake* [17] underpin this finding while arguing that with changing system goals the interpretation of an explanation should change to. The author also emphasizes that this requirement holds in different research fields like Psychology, Philosophy and AI. At the same time, *Leake* [18] uses the factors plausibility, relevance and usefulness for explanations concerning anomalies in regard to a given goal. Coming to the conclusion that "(m)any explanations can be generated for any event, and only some of them are plausible" [18]. The requirement we identify here is that a self-explanatory description must include not only regular information but also semantic information (about the meaning of the regular information) and context information for the context dependent meaning. This correlates with the overall goal of self-explanation proposed by *Müeller-Schloer et al.* [21] to enable systems to explain its current state, which seams to be impossible without providing contextual information.

4 Formal Foundation

Explanation of assumptions might informally be defined as a description to reveal the identity of some subject of interest. This might for example include information about its functionality. Imagine that we want do identify different boat types for tax reasons. We might not use the appearance to identify the difference of a rowing-, sailing- and a motor boat, because there might be different appearances in each class of boats. Instead, to identify the different boat classes, we need to describe some other details like the propulsion method and the tonnage of the boat. In contrast, if somebody wants to describe the different boat types to a child the functionality might be the detail separating the identities. In AI this fact is well known, since we seek different metrics to decrease intra class scatter and increase inter class scatter [7] (p. 121). Further, the explanation we must provide depends not only on the context but also on the reasoner how infers about it. With this in mind, an explanation should help the audience, to identify the classes a Subject of Interest SOI might be part of and with that better describe its identity to foster understanding of the explanation whereas the understanding determines the goodness of an explanation [17].

We now formalize these definitions, easing the creation of measurable properties of explanations to determine the quality of an explanations.

We define the amount of information transfered to the audience as a measure of quality of the explanation. First we want to define a domain as a set of information concerning this domain:

Definition 2. The information available in one domain $\mathbb{D}$ with $\mathbb{D} \subset \mathbb{I}$ and $\mathbb{I}$ bing all information available.

Here, the basic assumption we follow is, that in computer science where information is digitalized, information is a discrete entity. For example the chess move "Qxd4" (e.g. as move in the center game of a Danish Gambit) in the domain of playing chess is one piece of information $i \in \mathbb{I}$ in the domain of chess $\mathbb{D}_{chess}$. Where $\mathbb{I}$ is the amount of information available and $\mathbb{D}$ is the formal description of a domain as a proper subset of the information space $\mathbb{I}$. Consequently, a domain $\mathbb{D}$ contains those information necessary to create fully observable planing for the given domain. The following definition express what a reasoner is:

As illustrated in the boat example, the quality of explanations depends on the reasoner how infers about this explanation. Therefore we first need to define what a reasoner is.

Definition 3. A reasoner r is defined as an entity which includes new information $i \in \mathbb{I}$ into its knowledge-base $I_r \in \mathscr{D}$ where $\mathscr{D}$ is a σ-Algebra over $\mathbb{D}$.

This does not mean that all elements of $\mathbb{D}$ are available to each reasoner r. This offers the advantage that reasoners are able to infer in both fully and partial observable problems. To elude the problem of domain overarching knowledge, we define a domain as a σ-Algebra introducing the characteristic that all unions of information of one domain with information of the same domain is always part of the domain again. Later on we will utilize this and other characteristics of $\sigma - algebras$ to define a measure for explanations. Now, let $\mathscr{R}$ be a $\sigma - algebra$ of sets over all reasoners of concern where $r \in \mathscr{R}$. Further let $e \in \mathbb{E}$ be an explanation in some domain $\mathbb{D}$. Then we can define how an explanation maps to information by defining how the information in an explanation is transfered to the audience as follows:

Definition 4. $e \underset{r}{\rightarrow} i$ the explanation $e \in \mathbb{E}$ holds information $i \in \mathbb{I} \Leftrightarrow \exists r \in \mathbb{R}$ which is able to integrate i into its knowledge-base $I_r \in \mathscr{D}$ with the observation of e.

With this definition an explanation holds and transmits information to an audience if a reasoner of the audience can integrate new information into its knowledge-base. To avoid a philosophical discussion, we define that an explanation has to be understood by someone. Now that we have some definitions about a explanation, we will look at self-explanation to determine more specifically what exactly a explanation is.

The dictionary defines *self-explanatory* as: "easily understood from the information already given and not needing further explanation" [1]. This definition leads to the conclusion that the information given by self-explaining descriptions is sufficient for some reasoner in the audience to understand the SOI and that the explanation is given by the SOI. Taking this definition into account, we define a *degree of explanation* as follows:

Definition 5. Let $\mu : \mathscr{E} \rightarrow \bar{\mathbb{R}}$ be a measure with the σ-algebra $\mathscr{E}$ over $\mathbb{E}$ as some explanations to a affine extension of the real numbers $\bar{\mathbb{R}} := \mathbb{R} \cup \{+\infty, -\infty\}$ with $\mu(E) := \forall i \subseteq I \mid sup_{r \subseteq \mathbb{R}}(\sum_{\forall e \subseteq E : e \xrightarrow[r]{} i} (\delta_{i,i_r}))$. With $\delta_{i,i_r} = \begin{cases} 1, \text{ if } e \xrightarrow[r]{} i \text{ and } i_r \cup i \neq i_r \\ 0, \text{ if } e \xrightarrow[r]{} i \text{ and } i_r \cup i = i_r \\ -1, \text{ else} \end{cases}$

where $E \in \mathscr{E}$ is the set of explanation, $R \in \mathscr{R}$ is the audience observing the explanation and $I \in \mathscr{D}$ is the set of pieces of information which should be explained in this domain. We acknowledge, that this is a practical measure, since the degree of explanations drops when a explanation is repeated in front of the same audience several times. Further we chose the supremum instead of an average since for a scientific "proof of concept" we need one reasoner able to reason upon the explanation. With this definition of a measure for the degree of explanation, we can conclude that a theoretical complete self-explaining explanation with $\mu(E) = \mid \mathbb{D} \mid$ for some explanation $E \in \mathscr{E}$ could exist, so that no other explanation $e_i \in \mathscr{E}$ could explain the information i better to the audience. In practice such an explanation misses an example. But for a specific domain $\mathbb{D}$, an explanation e might be self-explanatory if the information space of $I_{r_1}, \ldots, I_{r_n} \in \mathscr{D}$ of the audience $r_1, \ldots, r_n, n \in \mathbb{N}$ of a domain d, is filled in that way, that the audience might reason to extract the entire information i hold by e by observing e.

On the one hand, the degree of self-explanation can be interpreted as the additional information needed to create understanding. On the other hand, as a measure depending on the reasoning capability of the audience and how the explanation fits to those capabilities. If no further information/capability is needed for some reasoner to understand the SOI, then the degree of self-explanation rises. The more information is needed the less the degree of self-explanation becomes, where in the worst case no useful information about the SOI can be extracted from the explanation.

In a domain, the information about the domain might be limited, and with that, the possibility for a good explanation might be given. To come back to our chess example the move: "Qxd4" probably needs further explanation. First we could explain the steno-notation syntax: The first element represent a chess piece here Q for the queen. The second element represents an optional action, here x which stands for making a capture and the last element *d4* concerning the location on the chess board where the move ends. Further we could additionally explain the meaning of "queen" or "making a capture". If e.g. the audience has watched the move of the chess game, the first explanation of the move given above has $\mu(E) = 0$. Under the assumption that the audience of this explanation does not know the steno-notation, the first explanation of the move given above could have $\mu(E) = -1$. Because there is one explanation and it is not understood. Now the second more detailed explanation can be of hight or lower quality. Since we have added multiple sub-explanations (Q,x,d4, queen and capture), if the audience still does not understand the explanation the measure of explanation can become, $\mu(E) = -6$. In this case all explanations did fail to transport information to the audience.

Further this explanation does not contain information about where the move started from, thus not being completely self-explanatory, since this depends on the context of the chess game. As we argued above, such contextual information is needed in self-explanatory descriptions for the effect of this example move. This reaches in the explanation of reasoning (the description of effect) and thus is out of scope of this work.

5 Conclusion

We can conclude that an explanation e transports information i to an audience of reasoners. The quality of an explanation can be measured by how much information the audience can extract from the explanation. So far, we define that an explanation becomes of higher quality if the degree of explanation rises. Our future work will be concerned with properties of explanation, in the attempt to make the definition more tangible. Further we want to integrate the existing structures of explanations like BDI and IOPE into explanations, to become able to represent an measure of self-explanation for existing descriptions.

References

1. Cambidge dictionary online (2012), `http://dictionary.cambridge.org/dictionary/british/self-explanatory?q=self-explanatory`
2. Braubach, L., Pokahr, A., Moldt, D.: Goal representation for bdi agent systems. In: Multi-Agent Systems, pp. 44–65 (2005)
3. Cheng, B.H.C., et al.: Software engineering for self-adaptive systems: A research roadmap. In: Cheng, B.H.C., de Lemos, R., Giese, H., Inverardi, P., Magee, J. (eds.) Software Engineering for Self-Adaptive Systems. LNCS, vol. 5525, pp. 1–26. Springer, Heidelberg (2009)
4. Chi, M.T.: Self-explaining expository texts: The dual processes of generating inferences and repairing mental models. In: Advances in Instructional Psychology, vol. 5, pp. 161–238. Routledge (2000)
5. Cole, W.G.: Understanding bayesian reasoning via graphical displays. SIGCHI Bull. 20(SI), 381–386 (1989), doi:10.1145/67450.67522
6. Druzdzel, M.J.: Qualitative verbal explanations in bayesian belief networks. Artificial Intelligence and Simulation of Behavior Quarterly 94, 43–54 (1996)
7. Duda, R.O., Stork, D.G., Hart, P.E.: Pattern classification and scene analysis. Part 1, Pattern classification, 2nd edn. Wiley (2000)
8. Elsaesser, C.: Explanation of probabilistic inference. In: Kanal, L.N., Levitt, T.S., Lemmer, J.F. (eds.) UAI, pp. 387–400. Elsevier (1987)
9. Grüninger, M., Hull, R., McIlraith, S.: A short overview of flows: A first-order logic ontology for web services. Data Engineering, 3 (2008)
10. Heckerman, D.E., Horvitz, E.J., Nathwani, B.N.: Toward normative expert systems: Part i. the pathfinder project. Methods of Information in Medicine 31, 90–105 (1992)

11. Hinchey, M.G., Sterrit, R.: Self-managing software. IEEE Computer 39(2), 107–109 (2006)
12. Jennings, N.R.: An agent-based approach for building complex software systems. Communications of the ACM 44(4), 35–41 (2001) (forthcoming)
13. Kaddoum, E., Raibulet, C., George, J.P., Picard, G., Gleizes, M.P.: Criteria for the evaluation of self-* systems. In: Proceedings of the 2010 ICSE Workshop on Software Engineering for Adaptive and Self-Managing Systems, SEAMS 2010, pp. 29–38. ACM, New York (2010), doi:10.1145/1808984.1808988
14. Kephart, J.O.: Autonomic computing: The first decade. In: Proceedings of the 8th ACM international conference on Autonomic Computing, ICAC 2011, pp. 1–2. ACM, New York (2011), http://doi.acm.org/10.1145/1998582.1998584, doi:10.1145/1998582.1998584
15. Kephart, J.O., Chess, D.M.: The vision of autonomic computing. Computer 36(1), 41–50 (2003), doi:http://dx.doi.org/10.1109/MC.2003.1160055
16. Klusch, M., Küster, U., Leger, A., Martin, D., Paolucci, M.: 4th international semantic service selection contest - performance evaluation of semantic service matchmakers (2010), http://www-ags.dfki.uni-sb.de/~klusch/s3/s3c-2010-summary-report-v2.pdf (last visited: November 1, 2013)
17. Leake, D.B.: Goal-based explanation evaluation. Cognitive Science 15(4), 509–545 (1991)
18. Leake, D.B.: Evaluating Explanations A Content Theory. Psychology Press (1992)
19. Martin, D., et al.: Bringing Semantics to Web Services: The OWL-S Approach. In: Cardoso, J., Sheth, A.P. (eds.) SWSWPC 2004. LNCS, vol. 3387, pp. 26–42. Springer, Heidelberg (2005), http://www.springerlink.com/index/rl5rlc8v64xvf0r8.pdf
20. Morris, C.: Foundations of the Theory of Signs, vol. 1. University of Chicago Press (1938)
21. Müller-Schloer, C.: Organic computing – on the feasibility of controlled emergence. In: Orailoglu, A., Chou, P.H. (eds.) Proceedings of the 2nd IEEE/ACM/IFIP International Conference on Hardware/Software CoDesign and System Synthesis, CODES+ISSS 2004, pp. 2–5. ACM, New York (2004)
22. Müller-Schloer, C., Schmeck, H.: Organic computing: A grand challenge for mastering complex systems. it – Information Technology 52(3), 135–141 (2010), doi:10.1524/itit.2010.0582
23. Salehie, M., Tahvildari, L.: Self-adaptive software: Landscape and research challenges. ACM Transactions on Autonomous and Adaptive Systems 4(2), 1–42 (2009), http://doi.acm.org/10.1145/1516533.1516538, doi:10.1145/1516533.1516538
24. Sooriamurthi, R., Leake, D.: Towards situated explanation. In: Proceedings of the Twelth National Conference on Artifical Intelligence, p. 1492 (1994)
25. Sycara, K., Klusch, M., Widoff, S., Lu, J.: Dynamic service matchmaking among agents in open information environments. SIGMOD Record 28, 47–53 (1999)

Semi-automated Generation of Semantic Service Descriptions

Nils Masuch, Philipp Brock, and Sahin Albayrak

Abstract. The increasing complexity and dynamics of distributed systems make the management and integration of new services more and more difficult. Automation processes for the definition, selection and composition of services for goal achievement can produce reliefs. However, a high degree of self-explanation of the services is obligatory for this. Today's multi-agent frameworks only provide insufficient solutions to this. Within this paper we will outline an approach, which enables the integration of semantic service descriptions into multi-agent systems with reasonable effort.

Keywords: OWL-S, Multi-Agent Framwork, Semantic Service Description, JIAC.

1 Introduction

In the past, the demand for modular, distributed and dynamic computer systems has increased rapidly. The reasons for this are manifold, ranging from maintainability and reliability to adaptability aspects, just to name a few [14]. In the field of multiagent systems (MAS) many of the current approaches try to account for these requirements. However, these systems usually are characterised by a high degree of complexity, which makes it difficult to provide a transparent way to define, lookup and invoke functionalities of software entities, such as agents. A promising approach to this is the service paradigm which leads to the development of service-oriented architectures (SOA) and is an ideal complement to multiagent systems [13]. One of its inherent strengths is typically the definition of a clear autonomy of each service, which means that it is represented as a separate module. Further, services are designed for enhancing the interoperability which is one of the key issues for distributed systems. Especially when talking about huge computer systems with different providers and parties involved these parameter are essential.

Nils Masuch · Philipp Brock · Sahin Albayrak
DAI-Labor, TU Berlin
Ernst-Reuter-Platz 7, 10587 Berlin, Germany
e-mail: nils.masuch@dai-labor.de

J.B. Pérez et al. (Eds.): *Trends in Prac. Appl. of Agents & Multiagent Syst.*, AISC 221, pp. 155–162.
DOI: 10.1007/978-3-319-00563-8_19

Beyond that, these systems also require mechanisms to adapt their behaviour to the environment or to current offers, since software services might be added or removed dynamically. In many approaches of current software systems the dynamic aspect is not regarded sufficiently. In order to encounter these challenges automated techniques that minimize the necessary intervention by developers are a promising approach [10]. One important issue in the automation process is the automated interpretation of services, allowing for a discovery of suitable services. Therefore the system has to be able to understand specific service parameters such as preconditions or effects. In practice a lot of so-called semantic service matcher components have been developed (e.g. [7], [9]), but, to the best of our knowledge, none of them has been fully integrated into a multi-agent framework. Doing so, the basic functionality is provided for the next step, which is the autonomic composition of services to reach a certain goal. The basic condition to develop these automation processes is a knowledge representation that enriches descriptions semantically, often described as ontologies. In practice the enrichment of real applications, which are more and more developed based on the agent-oriented programming paradigm (AOP), by ontologies is currently very troublesome if available at all. This means there is a clear lack of concepts combining the AOP world, which is widely used for programming, and the world of standardized formal semantics not only at runtime but also at design time.

In this paper we present a conceptual approach for integrating semantic service information into a MAS, taking the first step towards standard-based, dynamic and automatic service matching and composition systems.

In the remainder of this paper we will give some background information about the semantic service description language OWL-S. Thereupon we will introduce JIAC V (Java Intelligent Componentware, Version 5) [5], our own Java-based multi-agent development framework. Based on this framework, we will then present our approach of integrating semantic service descriptions into the development methodology of dynamic, distributed systems based on JIAC. Consequently we will present a use case, in which the dynamic invocation of services can be gainful. Finally, we will conclude with a short outlook of our next steps.

2 Background

In the context of semantic enrichment of services the semantic web community developed multiple approaches, such as SAWSDL[1], WSMO [2], hRests [3] and OWL-S [1] which are focusing on many identical aspects but are at the same time distinct in some relevant attributes. Due to the lack of space we are not able to discuss them thoroughly and refer to [8]. In the following we will focus on one of the most comprehensive ones, namely OWL-S, and describe it in more detail.

[1] `http://www.w3.org/2002/ws/sawsdl/`

2.1 OWL-S

OWL-S is based upon on the Web Ontology Language (OWL)[2], more precisely it is a specific OWL ontology, which is structured for describing service attributes. The ontology is split up into three parts, namely Service Profile, Service Grounding and Service Model.

The Service Profile describes all relevant parameters of the service for searching components. The most important attributes are the so-called IOPEs, which stands for Input, Output, Precondition and Effects. The Service Model enables the description of the service's processment. It defines how the service works, and if necessary which interactions will take place. For example a Service Model can combine multiple atomic processes to more complex components using constructs like loops or conditional expressions. The Service Grounding defines the invocation details of the service. The OWL-S standard leaves the kind of invocation technique open, allowing the integration of different transport protocols. For the definition of preconditions and effects OWL-S does not adhere to a specific rule language. One of the propositions to use is the Semantic Web Rule Language (SWRL) [6].

SWRL has been designed to formulate implication rules, which can be used on entities from OWL ontologies. A SWRL rule consists of two parts, the antecedent (body) and the consequent (head). For the evaluation of the rules the information from the head will be added to the knowledge base, if the conditions in the body are fulfilled. SWRL rules are comprised of any number of conjugated atoms. In order to evaluate the rules in acceptable time, the language must remain decidable. Therefore all variables which are required in the head part, have to be already present in the body part and further no complex class constructs may be used. Using SWRL according this way, it can be used together with OWL-DL and remains decidable.

2.2 JIAC V

Java **I**ntelligent **A**gent **C**omponentware V (JIAC V) is, as the name indicates, an open Java-based multi-agent framework currently being developed at Technical University of Berlin. The framework combines the agent-oriented with a service-oriented view. Services can either be invoked via explicit invocations of specific services or via an incomplete service description that leads to a service matching process returning the most appropriate one. Agents in JIAC V are located on Nodes. These nodes can be executed on different machines and are also responsible for middleware management and communication issues. For example each node manages the dynamics via a service- and an agent-directory according to FIPA Agent Management Specification[3].

The agents in JIAC V are component-based. Their structure can be very roughly divided into an execution cycle, a local memory and Agent Beans. The memory is a tuple space implementation, and provides parallel access of the data to all agent

[2] `http://www.w3.org/2004/OWL/`

[3] `http://www.fipa.org/specs/fipa00023/XC00023H.html`

components. The components also have the opportunity to sign up for listeners on memory data, in order to be informed about every change. The execution cycle is the heart of JIAC agents. This main thread processes existing events at regular intervals.

The different functionalities and behavioural rules of an agent are encapsulated in Agent Beans. These functions are defined as actions and can be made available in different scopes. Thus, an action can be defined as being visible to the agent itself, to all agents of one node or globally to all nodes.

One central component of JIAC V is the Semantic Service Matcher $SeMa^2$ [9], which enables the system to match OWL-S based service requests and advertisements. The matching process follows a hybrid approach, which considers both syntactic and semantic elements. For the evaluation of rules $SeMa^2$ utilises a OWL-DL Reasoner named Pellet [11], which is able to reason over SWRL constructs. The overall evaluation result is composed of multiple matching algorithms, namely Service Name Matching, Text Similarity Matching, Taxonomy Matching, Rule Structure Matching and Rule Reasoning. As a result the service matcher provides the requester with an assessment of all suitable service options in a ranking order.

3 Approach

As described within the previous chapter, OWL-S is a very comprehensive service desciption ontology with a huge flexibility in describing the IOPEs. However, this complexity leads to a significant drawback when it comes to the creation and integration into an application environment, since there are no sufficient tools available. In the following we propose our conceptual approach of integrating OWL-S into JIAC V by describing the different methodological steps towards a dynamic service platform.

The first essential step for an efficient interoperability is the definition of a domain ontology (see Figure 1). When developing project applications based on JIAC V often Eclipse Modeling Framework (EMF[4]) is being used. EMF comes with a direct transformation into Java. Further, there is an intensive research in transforming EMF models into OWL ontologies [12, 4]. Consequently, EMF is an ideal candidate to describe ontologies and transform them to the implementation language (Java) and the service description language (OWL / OWL-S).

As described in the JIAC V section, Java methods developed within Agent Beans can easily be declared as JIAC Actions by annotating the method heads. Figure 2 shows such a method head exemplarily. During runtime all relevant service information is being extracted and added to the agents service directories. Therefore the developer can fully concentrate on the implementation of the service's functionality.

However, what is missing here is the semantic enrichment of services for a detailed analysis of their functionality. Figure 3 shows a process diagramm illustrating the workflow of how semantical information should be enhanced to the core service data. At first the annotation head will be extended by a further tag named *semantify*, which specifies that the service should be descibed via an OWL-S ontology. Then,

[4] `http://www.eclipse.org/modeling/emf/`

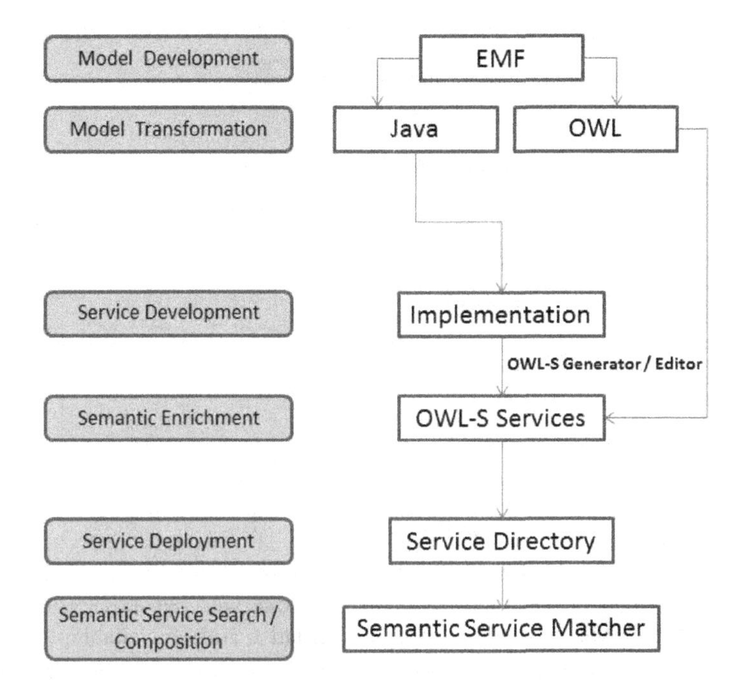

Fig. 1 Service integration methodology

```
@Expose(name = MobilityServiceBeanActions.ACTION_SEARCH_ROUTING_OPTIONS, scope = ActionScope.NODE)
public ArrayList<JourneyEvent> getJourneyOptions (Position location1, Position location2, Date time) {
```

Fig. 2 The JIAC V annotation mechanism

at compile time, the system checks whether a service description for this action has already been created. If not, a component named *OWL-S Generator* extracts all available information of the method head and the comment part (for now service name, service description, Java input params, Java output param) and maps the input and output parameters to the concepts of the corresponding ontology, which has to be defined as the reference ontology within the JIAC application settings. Consequently, a first OWL-S description will be created and loaded into an Eclipse based Editor. The developer is now able to modify and extend the service description (e.g. preconditions and effects). After the adaption phase the OWL-S description is stored persistently and a reference from the service to the description is being stored in the service header. At runtime the framework deploys and registers the service within the agent's service directory.

In order to find and integrate an appropriate service into a new application, the developer has to be able to specify the desired functionalities. Therefore there has to be some sort of Semantic Service Search Editor at design time, which is supporting the definitions of IOPEs and further restrictions, such as QoS parameters. The search

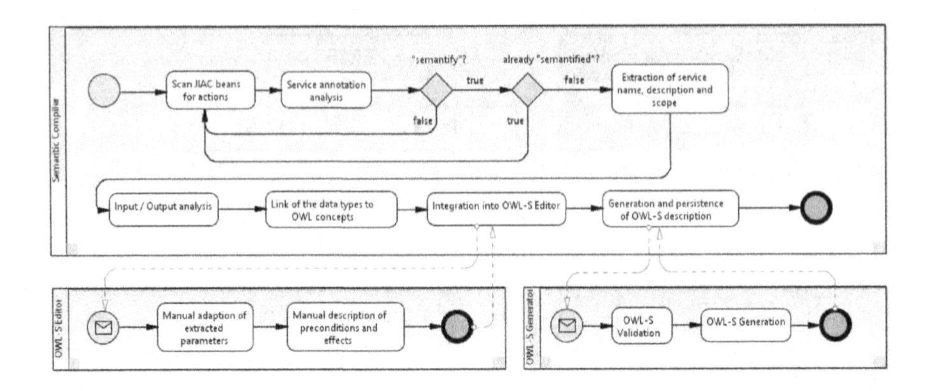

Fig. 3 Workflow of the OWL-S generation process

algorithm is then based upon the semantic service matcher SeMa^2 and returns all suitable and available services. The developer can select the most appropriate one, which is then hard coded into the invocation details. In a next step a hybrid approach is intended, in which the developer defines a specific service, but at times the service is not available, SeMa^2 is searching for an alternative and, if available, invokes it instead. Finally, the service matcher should be able to compose multiple services to reach a specific goal.

Our main intention with this approach is to enable the developer to easily utilise semantic service descriptions in order to have an increased flexibility and automation in highly dynamic application environments. In the next chapter we will identify a use case in which the additional overhead is justified.

4 Example

A domain that is well-suited for the integration of service automation techniques into huge computer systems is the field of mobility and transportation. Especially when thinking about urban traffic, people are asking for more flexible solutions regarding their personal needs. Therefore intermodal solutions or new approaches such as spontaneous car- or ride-sharing become more and more popular. Integrating different mobility providers into a distributed platform is an ideal use-case for an assistance system where dynamic aspects such as the availability of resources and context information (e.g. traffic jams) have to be considered for creating a user-specific solution. The combination of these offers can lead to composite services that support the user with intermodal and adaptable travel assistance. In this scenario services will be dynamically added or removed. Having a platform with comprehensive, but distributed information about mobility and routing options together with value added services offers the opportunity to develop a highly automatic service assistance planner. However, it requires an architecture which enables the developers to easily extend their services with semantical information, which is interpretable for standard-based matching and composition components. The concept described in the paper shall be a basic step towards this goal.

5 Conclusion

In this paper we presented a methodological and procedural approach of integrating semantic service descriptions into a multi-agent framework, in order to address the challenges of highly dynamic and complex application platforms. Our approach shows how OWL-S descriptions can be semi-automatically generated via standard Java method information and multi-agent framework specific annotations and integrated into the running system. Doing so, the software developer will be unburdened from declaring all service description information manually.

In a next step, we will implement the described concept within the JIAC V agent framework and upon that we aim to develop an open, intermodal mobility service platform as a first evaluation use case within a national government funded project.

References

1. Ankolekar, A., Burstein, M., Hobbs, J.R., Lassila, O., Martin, D.L., Mcilraith, S.A., Narayanan, S., Paolucci, M., Payne, T., Sycara, K., Zeng, H.: DAML-S: Semantic Markup for Web Services
2. Feier, C., Roman, D., Polleres, A., Domingue, J., Fensel, D.: Towards intelligent web services: The web service modeling ontology. In: International Conference on Intelligent Computing, ICIC (2005)
3. Gomadam, K., Vitvar, T.: hRESTS: an HTML Microformat for Describing RESTful Web Services. In: Proceedings of the 2008 IEEE/WIC/ACM International Conference on Web Intelligence, WI 2008 (2008)
4. Hillairet, G., Bertrand, F., Lafaye, J., et al.: Bridging EMF applications and RDF data sources. In: Proceedings of the 4th International Workshop on Semantic Web Enabled Software Engineering, SWESE (2008)
5. Hirsch, B., Konnerth, T., Heßler, A.: Merging Agents and Services — the JIAC Agent Platform. In: Bordini, R.H., Dastani, M., Dix, J., El Fallah Seghrouchni, A. (eds.) Multi-Agent Programming: Languages, Tools and Applications, pp. 159–185. Springer (2009)
6. Horrocks, I., Patel-Schneider, P., Boley, H., Tabet, S., Grosof, B., Dean, M., et al.: SWRL: A semantic web rule language combining OWL and RuleML. W3C Member Submission 21, 79 (2004)
7. Kapahnke, P., Klusch, M.: Adaptive Hybrid Selection of Semantic Services: The iSeM Matchmaker. In: Blake, M.B., Cabral, L., König-Ries, B., Küster, U., Martin, D. (eds.) Semantic Web Services, pp. 63–82. Springer, Heidelberg (2012)
8. Klusch, M.: Semantic web service description. In: Schumacher, M., Schuldt, H., Helin, H. (eds.) CASCOM: Intelligent Service Coordination in the Semantic Web. Whitestein Series in Software Agent Technologies and Autonomic Computing, pp. 31–57. Birkhäuser, Basel (2008)

9. Masuch, N., Hirsch, B., Burkhardt, M., Heßler, A., Albayrak, S.: SeMa^2: A Hybrid Semantic Service Matching Approach. In: Blake, M.B., Cabral, L., König-Ries, B., Küster, U., Martin, D. (eds.) Semantic Web Services, pp. 35–47. Springer, Heidelberg (2012)
10. Papazoglou, M.P., Traverso, P., Dustdar, S., Leymann, F.: Service-Oriented Computing: a Research Roadmap. Int. J. Cooperative Inf. Syst. 17(2), 223–255 (2008)
11. Parsia, B., Sirin, E.: Pellet: An OWL DL Reasoner. In: Proceedings of the International Workshop on Description Logics, p. 2003 (2004)
12. Rahmani, T., Oberle, D., Dahms, M.: An adjustable transformation from OWL to Ecore. In: Petriu, D.C., Rouquette, N., Haugen, Ø. (eds.) MODELS 2010, Part II. LNCS, vol. 6395, pp. 243–257. Springer, Heidelberg (2010)
13. Ribeiro, L., Barata, J., Colombo, A.: MAS and SOA: A Case Study Exploring Principles and Technologies to Support Self-Properties in Assembly Systems. In: Second IEEE International Conference on Self-Adaptive and Self-Organizing Systems Workshops, SASOW 2008, pp. 192–197 (2008)
14. Wooldridge, M.J.: An Introduction to MultiAgent Systems, 2nd edn. Wiley (2009)

Patience in Group Decision-Making with Emotional Agents

Denis Mušić

Abstract. Intense and usually stressful group decision-making has become a daily activity in the modern business environments which caused the emergence of interest in systems that allow simulation of group decision-making with agents as human representatives (surrogates). Development of representative agents is significantly enhanced through the use of methods that allow mapping of some of the most important human traits in the world of agents. These traits are emotions, personality and mood which gain importance by their direct effect on the process of individual and group decision-making. This paper presents the research results of applying the concept of patience to the emotional agents which should provide more stable and efficient group decision-making.

Keywords: emotional agents, patience, self-regulation coefficient.

1 Introduction

Development of representative human agents primarily requires an efficient way for mapping human traits into the world of agents. Recent research results have reported successful implementation of methods for mapping some of the most important traits like emotions, personality and mood. The enumerated properties of agents not only have great significance for the individual, but it is particularly interesting their impact on group decision making. Analysis of the modern business environments imposes the conclusion that there are a number of reasons why group decision making can result in failure and some of them are: the disturbed interpersonal relationships between participants, subjectivity, impatience, etc. These restrictions can be simulated by using multi-agent system that allows agents with emotions,

Denis Mušić
Universtiy "Dzemal Bijedic",
Faculty of Information Technology,
Mostar, Bosnia and Herzegovina
e-mail: denis@fit.ba

J.B. Pérez et al. (Eds.): *Trends in Prac. Appl. of Agents & Multiagent Syst.*, AISC 221, pp. 163–170.
DOI: 10.1007/978-3-319-00563-8_20

personality and mood to become human representatives in simulating real life meeting and produce a realistic meeting outcome.

Current studies are focused on creating an environment for intelligent interaction that can provide support for formal business meetings, tutorials, project meetings, discussion groups and ad-hoc interactions [1]. Context-aware emotion-based agent model presented in [2] ensures that clusters of agents bearing emotion-based features achieve agreements more quickly than those without such features. This model provides possibility to design intelligent agents with emotional awareness in order to simulate group decision-making processes. Aiming to improve the agent-based negotiation process, authors [3] have successfully incorporated affective characteristics such as emotions, personality and mood. Application of the aforementioned characteristics has also been presented in research [4] which introduced concept of virtual humans with a personality profile and real-time emotions and moods.

Model presented in this paper supports group decision-making realized by exchanging different types of arguments described in [5], where agents try to convince other participants to adopt their preferences. However, depending on the type of used arguments, the agents experience different kinds of positive and negative emotions that directly affect mood, and therefore the rest of the decision-making process [3].

The contribution of this work described in the following sections is integration of the concept of patience with the emotions as an important component for group decision making process. Introduction of patience is just one more step towards complete simulation of human qualities in order to assign owners' (human) characteristics to an agent which could make them realistic human representatives. By reducing intensity and undesirable emotional transitions, component of patience is entrusted with the task of acquiring the control over negative agent emotions and sudden emotional changes that can occur during group decision-making.

The paper is organized as following. Section 2 briefly describes methods for implementing personality, mood and emotions in world of agents. Section 3 introduces concept of patience and describes its integration with existing agent model. At the end, Section 4 presents model testing results aquired from multi-agent platform developed at GECAD[1].

2 Agents with Personality, Emotions and Mood

This paper is based on research described in [2] [3], according to which the agent structure is consisted of three layers: knowledge, reasoning and interaction layer. In this contex, Figure 1 plainly identifies position of patience as the newly proposed component inside agent structure.

[1] Knowledge Engineering and Decision Support Research Center, Polytechnic Institute of Porto.

2.1 Personality, Emotions and Mood

Psychology clearly defines interrelationship between emotions, personality and mood. Personality can be regarded as a set of predictable behavior by which people can be recognized and identified [6]. Personality types are identified by using well known FFM (Five Factor Model) model composed of a set of factors (dimensions) which describe specific personality traits that are rarely changed in the context of time and state [7]. Categorization of individual differences is carried out through the following five dimensions (called OCEAN): Openness, Conscientiousness, Extraversion, Amenity and Neuroticism.

Mood is defined as a feeling or a prolonged emotion that influences the complete state of the personality, and thus the perception of the environment [8]. Mood component presented in model from Figure 1 is in charge of generating agents mood based on the emotion intensity. Mood modelling is achieved by using PAD model [9] which consists of three dimensions: Pleasure (P), Arousal (A) and Dominance (D). These dimensions are fully independent, and can be presented in the PAD 3D space [10] as demostrated in [11] [12].

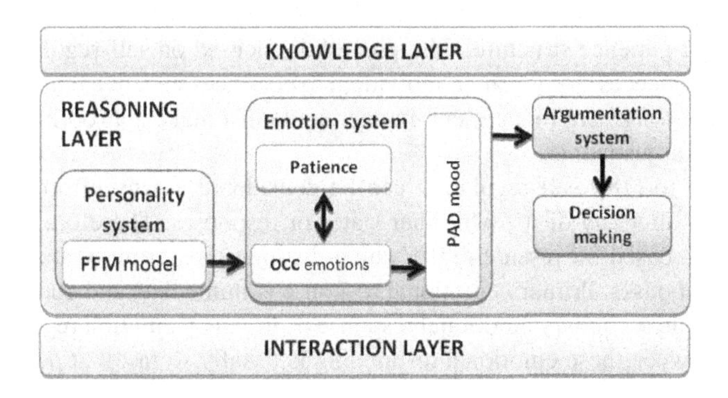

Fig. 1 Structure of the agents with emotions, mood, personality and patience

Emotions are considered as one of the indispensable components of the decision-making process. Agent emotions (Em) are defined as reactions to events, agents or objects, where the reaction is directly dependent on interpretation of certain situations [13]. Ortony, Clore and Collins presented the OCC model which describes total of twenty-two emotions that can be found in humans [14], after which they presented a shortened version of the original model [13] that contains the positive and negative emotions.

Emotions from OCC model can be mapped to the PAD space based on the values presented in [4]. Emotional system of agent presented in 1 provides a mechanism for selecting the most dominant emotions that have arisen as a reaction to some stimulus from the environment (event, action or object). Dominance is determined by the value of the difference between intensity and threshold (activation) of certain emotions. Considering its direct association with positive and negative reactions, the activation threshold value is obtained as the difference of Extraversion and Neuroticism of the OCEAN model.

As in real life meetings, agents with human traits such as emotion and mood can easily cause group decision-making process to fail. One of the reasons is appearance of negative emotions caused by actions of other agents which results in mood change and thus affects the argument selection process. Therefore, patience can be considered as an adaptation mechanism for group decision-making that will ensure its continuance in critical stages and reduce failures.

3 Patience within Emotional Agents

In everyday life, patience is considered the capacity to accept or tolerate trouble or delay without getting angry or upset. Very scarce number of papers deals with the narrow structure and methods of measuring patience. This can be confirmed by the fact that the model for measuring the patience presented in [15] is based on unpublished works that have set the fundamental theoretical structure of patience described in [16][17]. They proposed three mechanisms for understanding patient behavior: frustration-aversion, self-regulation, and temporal-altruism. Considering complexity of patience structure, this research is focused on self-regulation mechanism which is necessary when an individual experiences negative emotions (reaction), in order to behave in a patient manner, and must make appropriate cognitive or behavioral adjustments.

According to [18], self-regulation can be considered as any effort by the human being to alter any of its own inner states or responses. Therefore, integration of patience is based on research [19] which distinguishes primary and secondary emotional responses. Primary emotional response is immediate and completely unregulated, while secondary emotional response is driven by emotion regulation. The transition between these emotional responding is usually so fast that people hardly notice it, and therefore the focus of this research is on secondary emotional response.

In order to incorporate patience as the new component of emotional agents, we introduce the self-regulation coefficient (β) which is applied in two stages. The first stage of self-regulation coefficient application affects the intensity of negative emotions ($em \in Em^{-}$) as shown in Equation 1. The original Equation for calculating intensity of emotion is presented in [3] and based on the level of self-regulation capacity (SRC) described in a Equation 2, an agent is able to reduce the intensity of emotion.

$$I_{em} = \frac{\sqrt{P^2 + A^2 + D^2}}{\sqrt{3}} * \beta \quad (1)$$

As described in [20], strategies for measuring self-regulation have proliferated and can be divided into three categories: rating-scales, indices derived from behavior and personality inventories. In order to define the value of the self-regulation coefficient, we used a form of rating-scale category called Self-Regulation Questionnaire (SRQ) developed by [22]. Based on the SRQ score, we recoginzed three types of self-regulation capacities: High (β=0.65), Intermediate (β=0.75) and Low (β=0.85).

The values of self-regulation capacity presented in Equation 2 were obtained by experimentation, taking into account two stages of β application.

$$\beta = \begin{cases} 1 & \text{iff em} \in Em^{+} \\ 0.65 & \text{iff em} \in Em^{-}; I_{em} > activation; \text{SRC = High} \\ 0.75 & \text{iff em} \in Em^{-}; I_{em} > activation; \text{SRC = Intermediate} \\ 0.85 & \text{iff em} \in Em^{-}; I_{em} > activation; \text{SRC = Low} \end{cases} \quad (2)$$

Besides the self-regulation coefficient, one more component supports implementation of patience. The aforementioned component is based on the research [21] which reports that self-regulation of the secondary emotional response is implemented through monitoring and operating processes. During the monitoring process, an agent compares current state with a desired state and after that an operating process reduces any discrepancies between these two states. In this research we consider emotions of Tranquility as desired emotional states.

Algorithm 1. Alogrithm for reduction of discrepancies between emotions

$em_{new} \leftarrow$ monitor(Em)
$if \neg$ positiveEmotion(em_{new}) $then$
$remoteness \leftarrow$ emotionalDistance(em_{new},$em_{desired}$)
$em_{proposed} \leftarrow$ operationProcess(em_{new},$em_{desired}$, $remoteness, \beta$)
$Em \leftarrow Em \cup$ generateEmotion($em_{proposed}$)

The monitoring process is implemented, based on the distance or the correlation between certain emotions presented in [23] [24] [25]. Aforementioned researches have investigated the hypothesis that emotions that co-occur frequently within a certain period of time may be relatively accessible to one another; while those that co-occur infrequently may be less accessible. They constructed a remoteness index of emotional states that yielded shortest paths between different emotions. The remoteness score between two emotions, they argued, provides a quantifiable representation of how much emotion management it takes to move from one emotional state to another. Therefore, as presented in Algorithm 1, the remoteness index is used inside *emotionalDistance* for calculating distance between two emotions. Remoteness index describes distances between emotions and Dijkstra algorithm is used to find the shortest path between them. As an example of emotional transition we found that the shortest path from Tranquility to Distress includes transition through emotions of Joy and Fear., while the shortest path from Joy to Anger requires transition over emotion of Pride. After finding the shortest distance between two emotions, *operationProcess* is in charge of reducing discrepancies between a new (em_{new}) and a desired ($em_{desired}$) emotional state. Discrepancies between emotions are realized by using Equation $remoteness * SRC$ which shoud produce recommended emotional remoteness after supressing distance with β value. As an example of using Algorithm 1, we can use a situation when agent, with Intermediate level of SRC, experiences emotion of Distress . Since the path cost from the new emotion of Distress and the desired emotion of Tranquility is 12 and $\beta = 0.75$, affected path cost is closest to the emotion of Fear which should be generated by *generateEmotion* instead of Distress. In addition to the intensity of emotion, the impact of self-regulatory coefficient at the emotional transition represents the second phase of its implementation.

4 Model Evaluation

This research is based on the assumption that the component of patience combined with the component of experience, which is subject of our future work, will contribute faster reaching of an agreement. In the simulation environment, agents represent surrogates of the university professors during the meeting where they need to make a decision about the best student to participate in one of the world's most prestigious software development competitions. The easiest way to solve the aforementioned dilemma would be to select the student with the highest average. However, during selection process professors had to consider the students knowledge in different fields of science such as: System modelling, Databases, Programming, Web technologies and English language. In addition, each of the agents (professor) has different criterias when it comes to selecting the best student. the method which agents use for decision-making is known as MCDM (Multi-Criteria Decision Making). Based on the established criteria, this decision-making process allows to make a selection of alternatives that best match the goals and desires of the individual agent[26].

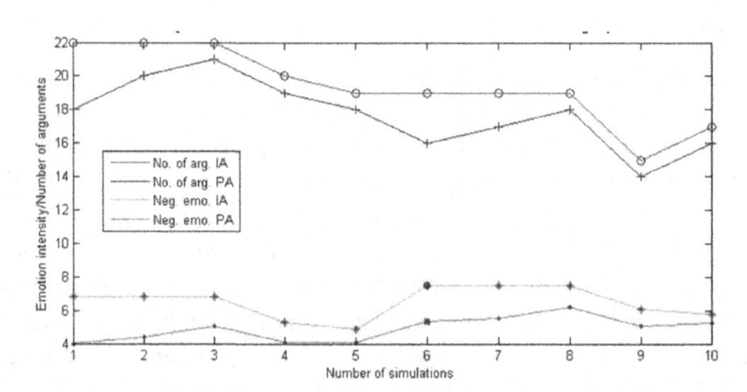

Fig. 2 Effect of applying patience on group decision-making with emotional agents

In order to fully evaluate the application of patience in group decision-making, we performed 10 simulations (SM) with four agents and each of them has different profiles (briefly described in Table 1). During the simulation, we observed the occurrence and intensity of negative emotions as well as their impact on the number of exchanged arguments in context of reaching an agreement. Previous research [3] has reported that the combination of agents with avoider and submissive personlaity type results in a reduced number of exchanged arguments while the number of exchanged arguments tends to increase with the combination of negotiator and aggressor agents. Therefore, the majority of the agents in the simulation had personality of negotiator and different values of the self-regulation capacity.

Table 1 Agent profiles

Simul.	Features	Agent1	Agent2	Agent3	Agent4
SM1	Personality	negotiator	negotiator	negotiator	aggressor
SM1	SRC	high	intermediate	low	high
....	....	....	....	....	....
SM10	Personality	aggressor	aggressor	aggressor	avoider
SM10	SRC	intermediate	intermediate	low	low

During the analysis, we summarized the intensity of negative emotions in each simulation. Number of exchanged arguments and intensity of negative emotions with patient (PA) and impatient agent (IA) during 10 simulations is presented in Figure 2. The results of model testing showed that patience is certainly desirable and implementable component in multi-agent group decision-making.

5 Conclusion and Future Work

Research results clearly confirmed that the introduction of the self-regulation coefficient and emotional transition regulation directly affects the reduction of negative emotions and number of exchanged arguments which should affect the speed of reaching an agreement within the group. However, one of the major limitations of the tested model is certainly a number of arguments that can be used within particular types of personality. This limitation prevents further testing of influence of the new coefficients in the specific situations of group decision-making. Also, we believe that component of patience will achieve its full potential in combination with component of experience which is subject of our future research. Component of experience will be realized by using a special form of Reinforcement learning called Q-learning in combination with Self-organizing neural network. Besides aforementioned, we plan to examine the relationship between self-regulation and the Neuroticism in order to affect emotional transitions.

Acknowledgements. Special thanks go to Carlos Ramos (Vice president of the Polytechnic Institute of Porto) and Goreti Marreiros (Knowledge Engineering and Decision Support Research Center-GECAD) on their generous assistance and provided opportunity to collaborate with them.

References

1. Tate, A., Chen-Burger, Y.-H., Dalton, J., Potter, S., Richardson, D., Stader, J., Wickler, G.: I-Room: a Virtual Space for Intelligent Interaction. IEEE Intelligent Systems 25 (2010)
2. Marreiros, G., Santos, R., Ramos, C., Neves, J.: Context-Aware Emotion-Based Model for Group Decision Making. IEEE Intelligent Systems 25 (2010)

3. Santos, R., Marreiros, G., Ramos, C., Neves, J., Bulas-Cruz, J.: Personality, Emotion, and Mood in Agent-Based Group Decision Making. IEEE Intelligent Systems (2011)
4. Gebhard, P.: ALMA: A Layered Model of Affect. In: AAMAS. ACM Press (2005)
5. Kraus, S., Sycara, K., Evenchik, A.: Reaching Agreements Through Argumentation: A Logical Model and Implementation. Artificial Intelligence 104 (1998)
6. Ryckman, R.M:Theories of Personality. Wadsworth Publishing Company (2003) ISBN: 978-0534619886
7. Howard, P.J., Howard, J.M.: The Big Five Quickstart: An Introduction to the Five-Factor Model of Personality for Human Resource Professionals. Education Resource Information Center (1995)
8. Clark, A.V.: Psychology Of Moods. Nova Publishers (2005) ISBN: 978-1594543098
9. Mehrabian, A.: Analysis of the big-five personality factors in terms of the PAD temperament model. Australian Journal of Psychology 48 (1996)
10. Frijda, N.H.: The Laws of Emotion. Lawrence Erlbaum Associates (2006) ISBN: 978-0805825978
11. Arellano, D., Varona, J., Perales, F.J.: Generation and visualization of emotional states in virtual characters. Computer Animation and Virtual Worlds 19 (2008)
12. van der Lei, K.: A Review of the Current State of Emotion Modeling for Virtual Agents. In: 14 Twente Student Conference on IT (2011)
13. Ortony, A.: On Making Believable Emotional Agents Believable. Emotions in Humans and Artefacts. MIT Press (2003)
14. Ortony, A., Clore, G.L., Collins, A.: The Cognitive Structure of Emotions. Cambridge Universtiy Press (1988)
15. Dudley, K.C.: Empirical Development of a Scale of Patience. Dissertation submitted to the College of Human Resources and Education at West Virginia University (2003)
16. Blount, S., Janicik, G.A.: Comparing social accounts of patience and impatience. University of Chicago (1999) (unpublished manuscript)
17. Blount, S., Janicik, G.A.: What makes us patient? The role of emotion in socio-temporal evaluation. University of Chicago (2000) (unpublished manuscript)
18. Baumeister, R.F., Vohs, K.D.: Handbook of self-regulation: Research, theory, and applications. Guilford Press, New York (2004)
19. Lazarus, R.S.: Progress on a cognitive-motivational-relational theory of emotion. American Psychologist 46(8), 819–834 (1991)
20. Hoyle, R.H., Bradfield, E.K.: Measurement and Modeling of Self-Regulation: Is Standardization a Reasonable Goal? In: National Research Council Workshop on Advancing Social Science Theory (2010)
21. Carver, C.S., Scheier, M.F.: On the Self-Regulation of Behavior. Cambridge University Press (2001) ISBN: 978-0521000994
22. Brown, J.M., Miller, W.R., Lawendowski, L.A.: The Self-Regulation Questionnaire
23. Lively, K.J., Heise, D.R.: Sociological Realms of Emotional Experience. American Journal of Sociology 109 (2004)
24. Lively, K.: Emotional Segues and the Management of Emotion by Women and Men. Social Forces 87, 911–936 (2008)
25. Morgan, R., Heise: Structure of Emotions. Social Psychology 51(1), 19–31 (1988)
26. Triantaphyllou, E.: Multi-Criteria Decision Making Methods: A comparative Study. Applied Optimization, vol. 44. Springer (2000) ISBN: 978-0792366072

Analysis of Web Usage Data for Clustering Based Recommender System

Shafiq Alam, Gillian Dobbie, Patricia Riddle, and Yun Sing Koh

Abstract. Implicit web usage data is sparse and noisy and cannot be used for usage clustering unless passed through a sophisticated pre-processing phase. In this paper we propose a systematic way to analyze and preprocess the web usage data so that data clustering can be applied effectively to extract similar groups of user. We split the entire process into analysis, preprocessing and outlier detection and show the effect of each phase on Java Application Programming Interface (API) documentation usage data that is collected from our server logs. We use the extracted clusters for web based recommender systems and present the accuracy of the recommendations.

1 Introduction

Web mining, a sub domain of data mining, uses standard Knowledge Discovery (KDD) techniques to extract patterns from web data. Web data is composed of the content, structure and usage of the web pages. Web structure mining aims to find patterns in the structure of web pages. Web content mining deals with the contents of web pages and Web Usage Mining (WUM) explores patterns in web usage data [10].

Understanding the web usage data is very important to comprehend the behavior of individual web-users, who usually follow a particular sequence while browsing the web. Discovery of such patterns is the ultimate goal of WUM. One of the recent applications of web usage patterns are recommender systems. Recommender systems are tools that assist users to find a particular resource based on prior knowledge about the behavior and usage of similar users and resources. Recommender systems based on implicit data; where users do not actively participate, and the data is

Shafiq Alam · Gillian Dobbie · Patricia Riddle · Yun Sing Koh
Department of Computer Science, University of Auckland,
38 Princes Street, 1010, Auckland, New Zealand
e-mail: sala038@aucklanduni.ac.nz,
{gill,pat,ykoh}@cs.auckland.ac.nz

J.B. Pérez et al. (Eds.): *Trends in Prac. Appl. of Agents & Multiagent Syst.*, AISC 221, pp. 171–179.
DOI: 10.1007/978-3-319-00563-8_21

gathered implicitly from the usage logs is a typical example of WUM based recommender systems. On the one hand, development of implicit recommender systems need huge amounts of data about users' interests while on the other hand the size of the data causes inefficiencies in near real time pattern extraction. Studies have shown that nearly 60% of web usage data contains irrelevant information, which includes image files which are automatically generated with page hits, web bots' and crawlers' requests which are machine generated activities, and some document formatting information. The analysis of web usage data and the subsequent preprocessing of the data are important tasks that need to be carried out before the actual pattern discovery starts. In this paper we present a systematic way of analysing and preprocessing the data prior to pattern extraction. As a case study, we examined web usage data from the University of Auckland, Department of Computer Science web log (CS-WebLog), which contains users' requests to access the Java Application Programming Interface (API) documentation.

2 Related Work

Research work in web usage analysis and preprocessing has significantly improved the quality of the patterns generated from web usage data. However, there has been little research on the analysis and preprocessing of weblogs. Castellano et al. [6] proposed a weblog data preprocessor which generates sessions of web users' from the user requests in the log, whereas, [12, 13] focus on detailed weblog analysis and preprocessing for mining association rules from web usage data. Cooley et al. [8] suggested different approach by using user identification, session identification, path completion and formatting to enable the weblog to be used for mining purposes. Another work in the area of web usage data analysis and cleaning is in [2, 3] where preliminary analysis and preprocessing is carried out for web usage data clustering and recommender systems. Previous reported work lacks both systematic analysis of the web usage log and a phase-by-phase transparent preprocessing approach. The contribution of this paper is twofold. We propose a step-by-step procedure for analysis and preprocessing for WUM-based recommender systems, and present a practical scenario of web usage data cleaning. Due to the page limitation we only present the results which are most relevant. The following two issues are central to the proposal of the analysis approach. (1) Presenting a case study of real world raw data that is large enough to be used for WUM based recommender systems, and (2) splitting of the cleaning process into sub areas such as analysis, preprocessing, and outlier detection.

3 Web Usage Data Analysis

Web usage data comes from web server logs, cache servers, and cookies. Web logs are the main repositories which provide users' activity data for WUM in Common Log Format (CLF). Log data needs to be passed through different preprocessing steps such as selection of relevant attributes, privacy preservation, cleaning and transformation. Primary attributes of web users can directly be obtained from the log such as IP address, date, time, page requested, page size, response and referrer. Secondary attributes such as user visit, sessions, session length, episode, sequence of navigation, and semantic information are extracted by manipulating the primary attributes and web documents. For details of WUM preprocessing one can refer to [12]. We divide the preprocessing for WUM into three sub-phases, analysis of the log, general preprocessing and algorithm specific preprocessing.

Our experimental data is composed on Java API documentation usage from the Department of Computer Science's web logs containing raw requests from 2006 to 2010, comprising 5.2 million Java API requests. We pass this data through a sophisticated preprocessing phase to extract usage sessions. The preprocessing method with preliminary results can be found in [2] and [3]. Analysis of data provides information about population size, number of potential patterns, and applicability of data mining techniques to the data and helps users to plan which pattern extraction method should be used. Simple statistics such as number of data vectors, number of attributes of each data vector, number of users, number of pages, mean and standard deviations for different measures are calculated in this phase. In the case of the Java API usage data, analysis identifies which part of the API is of interest for building a recommender system. For our particular case we divide the analysis phase into three phases. **Pre-analysis preparation** collects simple statistics, tokenizes the requests, analyzes the broken request strings, and fuses different logs together into one log. In our case the raw data provided was time stamped and divided into monthly and weekly data. We tokenize all the CS-WebLog data, remove all broken strings, and fuse it into one single log. **Log-based analysis** provides information about the overall structure of the log including total requests in the log, requesters and their requests, distinct IP's and requests, image requests, and CSS requests. This phase can be performed without designing sophisticated queries. Once the web log is analyzed as a whole, the **subject-based analysis** phase starts which targets information of different subject areas in the web logs such as requester (distinct, active, non-active IP), page (distinct pages, active page, non-active page), and time (year, semester, month, day, and time).

Web users with very low numbers of requests do not contribute much towards pattern generation. We found that out of 70,000 different IPs, 60% of them have 4 or fewer requests and their requests are quite diverse so their contribution is very small, so we removed such requests. Web pages which have fewer requests, also do not contribute to the pattern discovery process. So we pruned all pages that have been browsed by fewer than 10 users. Another important entity of usage data is the time domain. The entire time span of our data is broken down into years, months, weeks and days. Each of these time spans was then analyzed separately to see how they relate to each other. For example all weekends have low request hits while working days have a high number of requests. Figure 1(a) shows the number of requests for Java API documentation by date. Spikes show the peak dates of the academic calendar. Figure 1(b) shows a month of usage (Dec-Jan 2007) of the API

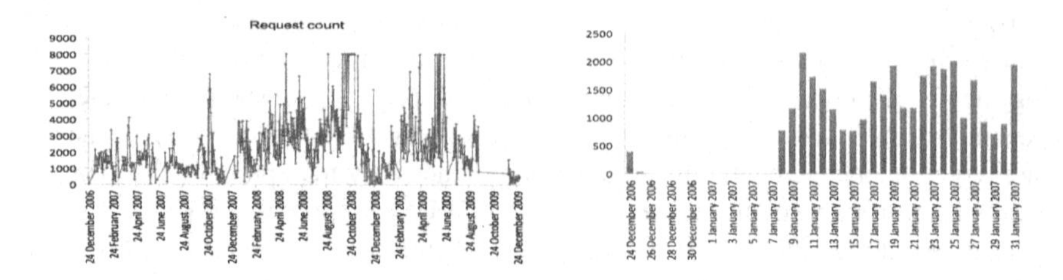

Fig. 1 Overall request count against dates and one month analysis (a) Number of requests from 2006 to 2009 (b) Number of requests for Dec-Jan 2007

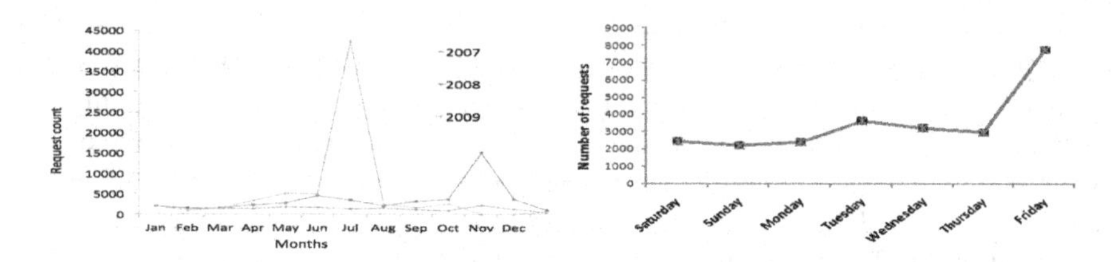

Fig. 2 Monthly analysis and daily average (a) Page requests per month (b) Page requests per week

documentation depicting the holiday seasons. Figure 2(a) shows average requests per months highlighting the trend amongst different months of the year. Figure 2(b) shows the average requests per week.

4 Data Preprocessing for Recommender Systems

We divided the preprocessing for recommender system into two phases. In **general processing**, the data is anonymized, filtered and transformed into relational database tables while pruning of irrelevant data takes place during algorithm **specific preprocessing**. Below we explain the activities of each preprocessing phase.

General preprocessing: General preprocessing transfers the data from a raw format to relatively cleaner data. It does not necessarily clean the data enough to be directly used by every data mining technique as each data mining algorithm has its own requirements. The steps of general preprocessing are listed below.

- *Privacy preserving :* It deals with anonymity, hiding the identity and personal information of the users. The aim of this step is to avoid compromising sensitive information about the user's interest and ensure his privacy.
- *Filtration :* Filtration is the process by which useful data vectors are extracted from the repository and non-helpful requests are removed, such as already known web robot requests and image file requests.

- *Transformation:* For effective querying of the data, data is mapped into a relational database. In the usage database semantic concepts are stored and divided into different but dependent concepts. The relationship between these concepts is used to compute secondary information.

To extract web users' secondary attributes, the log has to go through some manipulation such as finding session, semantic topic, visit, and episode information. These operations are performed in algorithm specific preprocessing. Grouping the primary data and secondary data gives us a format where queries can be directly applied and data can be mined for patterns.

Algorithm Specific Preprocessing: Algorithm specific preprocessing prepares the data for an individual mining algorithm i.e. for clustering, classification rule mining, and association rule mining. Further details of algorithmic specific preprocessing on a case study from NASA weblog can be found in [2] and [3]. Below we outline some of the important tasks of algorithm specific preprocessing.

- *Semantic identification:* Semantic analysis deals with finding content similarity by using different similarities such as text similarities and graph similarities. In our preprocessing we did not perform semantic identification and only considered the sequence and visit time similarities of the users.
- *Image request treatment:* It identifies the image requests and treat them appropriately. We removed all the image requests during the preprocessing phase.
- *Secondary attribute extraction:* Secondary attributes extracts attributes such as sequence of visits, session information, and episode information. We extract secondary attributes such as session length, pages per session, and amount of data per session from the web log.
- *Aggregation of attributes to form data vectors:* In this phase the attributes about the data are aggregated and put in a format which gives sufficient information about an entity.

Overall the analysis of web logs gives an overview of the visit distribution of the web users. Once the data passes through preprocessing and mining, there can be two major issues with the extracted patterns. Firstly, whether the extracted patterns are useful, and secondly, can the patterns be interpreted correctly and efficiently to be used in a real time application. For instance, for an application area, clustering based recommender system; these problems can be as follows. Are the clusters generated accurate, and can they be used for a recommender system? To solve the first problem, different clustering validation measures such as intra-cluster distance, inter cluster distance, and clustering indices could be used. These measures are greatly affected by noise, classification error, and outliers in the data.

Outlier Detection for Recommender Systems: While building explicit recommender systems, noise and web bots generated data skews the pattern extraction process. Although there are some known crawlers, i.e. http://www.iplists.com/, but the list is out of date and there are a number of bad bots that are unknown. Outlier detection can be used to detect web-bots based on their features rather than relying on the provided list. We used our HPSO-clustering based outlier detection method [5]

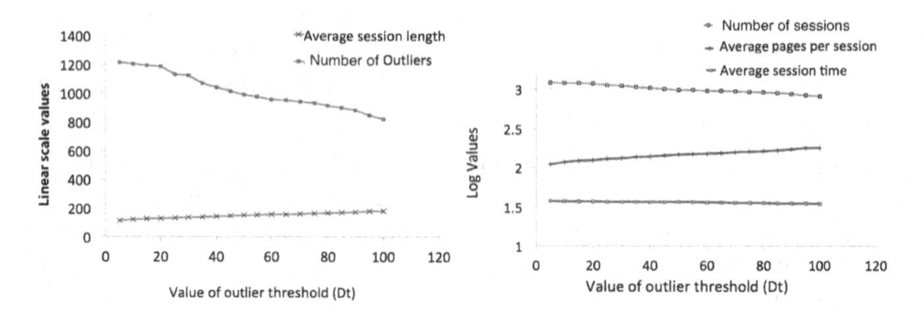

Fig. 3 (a) Distance threshold and length of sessions (b) Distance threshold vs number of outlier session, average length of outlier session, and average time of outlier session (log-scaled values)

to detect the web-bots and removed them from the data to be used for recommender systems. For detecting sessions from web-bots, we followed the same experimental steps as we followed for outlier detection with the same parameters [5]. Figure 3(a) shows the number of detected outliers and the average length of the session (pages per session) against different threshold values of D_t (the distance threshold value). Figure 3(b), shows the relationship between the value of D_t and the number of suspected outliers, number of pages per session, and average amount of time per session. For larger values of D_t the average session length is lower and pages per session is higher. This shows that the detected web-bots visit a large number of pages in a smaller amount of time. For a genuine web user, the average number of pages per session is 28 pages, while the suspected outliers identified by our method have 180 pages per session. Similarly, the average session length for genuine users is 28 minutes, while for the suspected outliers it is 36 minutes. **Denoising and Duplicating:** After the preprocessing phase data is relatively cleaner but sparseness could not be eliminated completely due to the large number of distinct pages in Java API. The analysis shows that almost 90% of the total data goes into only two of the clusters. Due to space limitations, we omit the figure showing these clusters. With such clusters the goals of accurate clustering based recommender systems cannot be achieved as highly populated clusters with many data elements per cluster cause inefficiencies in generating recommendations. To alleviate the problem of sparsity we removed those API pages which have fewer hits received during a specific time period (month), and termed this process as denoising. Figure 4 shows the results when a threshold of 20 pages was selected. In Figure 4, the data is distributed into many clusters which have clear boundaries between different clusters and can be used for generating recommendations. The CS-WebLog data lacks density of similar sessions so the intra-cluster distance before denoising across the generation was relatively flat. This means that there is no difference in the intra-cluster distance in subsequent generations. To overcome this problem we duplicated the data to mimic a denser problem space, which could provide enough patterns for recommender systems. We performed a complete duplication of the data, which artificially boost the

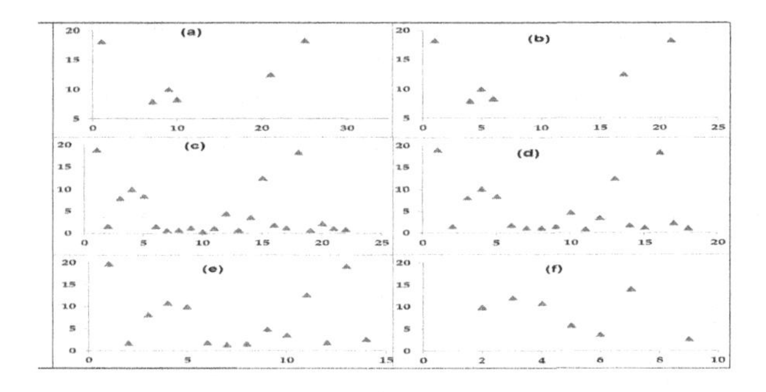

Fig. 4 Percentage of cluster membership after denoising (a) 30 clusters (b) 27 clusters (c) 23 clusters (d) 19 clusters (e) 15 clusters (f) 11 clusters

Position	Top 5 Recommendations		Top 10 Recommendations	
	Average	Std. Dev.	Average	Std. Dev.
Centroid	0.65	0.08	0.38	0.09
Random	0.58	0.08	0.48	0.06
Low Populated	0.66	0.08	0.52	0.1

Recommendations Precision (Top 5)				
	Cluster 1		Cluster 2	
Recommendation	Average	Std. Dev	Average	Std. Dev
3rd	0.80	0.10	0.90	0.20
4th	0.61	0.07	0.92	0.13
5th	0.52	0.08	0.94	0.10
6th	0.47	0.11	0.90	0.15
Average	0.60	0.09	0.91	0.04

Fig. 5 Precision of our proposed clustering based recommender system

clusters, however it enabled the generation of well-defined clusters for the recommender system. Complete duplication does not significantly change the clustering configuration of the data but it increases density in each cluster, which is one of the requirements of recommender systems.

5 Recommender System: An Application

After analysis and preprocessing, different data mining techniques such as clustering can be applied on the data. The extracted clusters can then be used to build a knowledge base for recommender systems. A number of recommender systems based on WUM have been proposed [7] [9] [11]. To generate recommendations, we analyzed, preprocessed web usage data, and removed web bot from the data. In the second phase the cleaned data was clustered using our proposed clustering technique. For an active user, recommendations based on nearest neighbours were generated and ranked based on the distance of the neighbours to that active user. Figure 5 shows the precision of the proposed recommender system based on the analysis presented in this paper. More details of modelling and experimentation of our proposed recommender system are given in [1] [4]. The average precision achieved is between 60% to 70%, while in some cases we achieved more than 90% precision. Our recommender system takes Java API usage requests as an input and passes them through the analysis phase to identify a region of interest, performs both preprocessing steps and transforms the data into a database. The preprocessed data is then used for clustering of usage behavior. An active user can now be compared with these clusters. Depending on the cluster to which the user belongs, a number of recommendations are generated for that user. Web usage clustering for recommendation reduces the problem space and increase the efficiency of generating recommendations and filtering based on the distance of the active user's session to the neighborhood. The same distance measure is also used to rank the recommendations.

6 Conclusion

Analysis of data is very important to gain the domain knowledge. We propose a step-by-step process of analysis for Java API usage mining and results are explained where needed. We also include some statistics about general preprocessing and algorithm specific preprocessing required for recommender systems. We aimed to focus our analysis on our overall goal; to build a recommender system which guides the user of the Java API documentation to a particular element of the API documentation. One of our future goals is to provide clean, and focused web usage data to be used as a benchmark for testing implicit recommender systems.

References

1. Alam, S.: Intelligent web usage clustering based recommender system. In: Proceedings of the Fifth ACM Conference on Recommender Systems, pp. 367–370. ACM (2011)
2. Alam, S., Dobbie, G., Riddle, P.: Particle swarm optimization based clustering of web usage data. In: Proceedings of the 2008 IEEE/WIC/ACM International Conference on Web Intelligence and Intelligent Agent Technology, pp. 451–454. IEEE Computer Society (2008)
3. Alam, S., Dobbie, G., Riddle, P.: Exploiting swarm behaviour of simple agents for clustering web users' session data. In: Cao, L. (ed.) Data Mining and Multi-agent Integration, pp. 61–75. Springer US (2009)
4. Alam, S., Dobbie, G., Riddle, P., Koh, Y.: Hierarchical pso clustering based recommender system. In: 2012 IEEE Congress on Evolutionary Computation (CEC), pp. 1–8. IEEE (2012)
5. Alam, S., Dobbie, G., Riddle, P., Naeem, M.: A swarm intelligence based clustering approach for outlier detection. In: IEEE Congress on Evolutionary Computation (CEC), pp. 1–7 (2010), doi:10.1109/CEC.2010.5586152
6. Castellano, G., Fanelli, A.M., Torsello, M.A.: Lodap: a log data preprocessor for mining web browsing patterns. In: Proceedings of the 6th WSEAS Int. Conf. on Artificial Intelligence, Knowledge Engineering and Data Bases, vol. 6, pp. 12–17. World Scientific and Engineering Academy and Society (WSEAS), Stevens Point (2007)
7. Castro-Herrera, C., Duan, C., Cleland-Huang, J., Mobasher, B.: Using data mining and recommender systems to facilitate large-scale, open, and inclusive requirements elicitation processes. In: Proceedings of the 2008 16th IEEE International Requirements Engineering Conference, pp. 165–168. IEEE Computer Society, Washington, DC (2008), doi:10.1109/RE.2008.47
8. Cooley, R., Mobasher, B., Srivastava, J., et al.: Data preparation for mining world wide web browsing patterns. Knowledge and Information Systems 1(1), 5–32 (1999)

9. Gemmell, J., Shepitsen, A., Mobasher, M., Burke, R.: Personalization in Folksonomies Based on Tag Clustering. In: Proceedings of the 6th Workshop on Intelligent Techniques for Web Personalization and Recommender Systems (2008)
10. Mobasher, B.: Web mining overview. In: Encyclopedia of Data Warehousing and Mining, pp. 2085–2089. IGI Global (2009)
11. Sarwar, B., Karypis, G., Konstan, J., Riedl, J.: Recommender systems for large-scale e-commerce: Scalable neighborhood formation using clustering. In: Proceedings of the Fifth International Conference on Computer and Information Technology (2002)
12. Tanasa, D.: Web usage mining: Contributions to intersites logs preprocessing and sequential pattern extraction with low support. PhD thesis (2005)
13. Tanasa, D., Trousse, B.: Advanced data preprocessing for intersites web usage mining. IEEE Intelligent Systems 19, 59–65 (2004)

Multi-label Classification for Recommender Systems

Dolly Carrillo, Vivian F. López, and María N. Moreno

Abstract. Multi-label classification groups a set of supervised learning methods producing models capable of classifying examples in more than one class. These methods have been applied in diverse fields; however, the field of recommender systems has been hardly explored. In this work, books' recommendation data are used to evaluate the behavior of the main multi-label classification methods in this application domain. The experiments carried out demonstrated their suitability to provide reliable recommendations and to avoid the grey sheep problem

Keywords: Web Mining, Recommender Systems, multi-label classification.

1 Introduction

Providing users with individualized suggestions about goods and services by means of recommender systems is a form of web system personalization. The web sites offering good recommendations attract and retain more clients than traditional ones. Recommender systems can be categorized in two main approaches: content-based and collaborative filtering. In content-based methods web items are recommended to a user based on those he has been interested in the past. On the other hand, collaborative filtering methods are not only based on preferences of the user who is the target of the recommendation but they are also based on opinions about items given by other users (neighbors) who have similar preferences.

The data mining techniques are propitious for this kind of systems since they allow to discover patterns in big collections of information, which can be useful for inducing user profiles, correlation between items, prediction of preferences and so on. With this purpose, diverse data mining methods have been investigated in the application domain of recommender systems. In this context, these techniques

Dolly Carrillo · Vivian F. López · María N. Moreno
Dept. of Computing and Automatic, University of Salamanca,
Plaza de los Caídos s/n, 37008 Salamanca
e-mail: {dolly.carrillo,vivian,mmg}@usal.es

J.B. Pérez et al. (Eds.): *Trends in Prac. Appl. of Agents & Multiagent Syst.*, AISC 221, pp. 181–188.
DOI: 10.1007/978-3-319-00563-8_22

are categorized as *Web Mining* methods, because they must be adapted in order to process Web data. Some of them, used for building classifications models, are association rules, neural networks, decision trees and Bayesian networks. The main methods of multi-label classification combine the above mentioned techniques and they have demonstrated to be efficient in a wide variety of application fields but their use in the recommender system domain is scarce. The aim of this work is to explore this area and make a contribution to relieve this lack. Due to the fact that multi-label classification methods are used to produce models capable of assigning a set of labels to new examples, their application might be of great interest to classify a user in one o more profiles and to predict the ratings he could give to some products in order to recommend them to him according to these predictions. In this way, the grey sheep problem, which is presented by many recommender systems, can be avoided.

The rest of the paper is organized as follows: Sections 2, 3 and 4 includes the fundamentals of multi-label classification and recommender systems and some related works. In section 5 the empirical study carried out with the *BookCrossing* dataset is presented and finally the conclusions are given in section 6.

2 Multi-label Classification

Nowadays multi-label classification methods are needed in multiple applications such as classification of proteins, music categorization, semantic classification of scenes or classification of educational content in e-learning environments. In all these applications the objects to be classified can belong to more than one class. For example, a photography can represent more than one concept as late afternoon and beach at the same time. In the same way, several labels can be assigned to a learning object and their characteristics can be used to improve the experience of students and teachers in the search and classification of educational contents. Multiple similar cases can be found in the literature.

For a formal description of multi-label classification task, $L = \{\lambda_j : j = 1...q\}$ can be used to denote the finite set of labels and $D = \{ (x_i; Y_i), i = 1...m\}$ to denote the set of training instances, where x_i is the vector of characteristics and $Y_i \subset L$ is the subset of labels of the instance i.

The subset of labels of the instance i is defined then as a binary vector, $Y_i = \{y_1, y_2...y_q\}$, where every $y_j = 1$ indicates the presence of the label λ_j in the set of relevant labels for x_i. Using this convention, the output space can be also defined as $Y = \{0, 1\}^q$. Finally, the aim of the classification multi-labels it is to assign the correct set of labels for a new instance x, that is, to predict if the label λ_j must be assigned or not to the example x.

There exist two main tasks in supervised learning from multi-labeled information [21]: multi-label classification (MLC) and label ranking (LR). MLC refers to learning models providing as output a bipartition of the label set into relevant and irrelevant labels, however LR involves learning models providing labels ordered according to their relevance for a given instance.

Multi-label classification and ranking methods can be arranged in two approaches [21]: *Problem transformation methods*, and *algorithm adaptation methods*. The first group of methods are algorithm independent. They transform the multi-label classification task into one or more single-label classification, regression or label ranking tasks. Some methods in this group are: *Binary relevance* (BR)[23], *Label Powerset* (LP), *ranking by pairwise comparison* (RPC) [7], *multi-label pairwise perceptron* (MLPP) [10] and *Calibrated label ranking* (CLR) [5]. Algorithm adaptation methods extend specific learning algorithms in order to handle multi-label data directly. There are several proposals for different learning techniques:

- Decision trees and Boosting. C4.5 [4], AdaBoost.MH and AdaBoost.MR, have been adapted for the treatment of multi-labeled data.
- Probabilistic methods. This approach encloses proposals as [9] Parametric Mixture Models (PMMs)[24] and Conditional Random Field (CRF) [6].
- Neural networks and support Vector Machines (SVM). The most representative algorithms in this category are Back-Propagation Multi-Label Learning (BPMLL) [28], Radial Basis Function Neural Networks for Multi-Label Learning (ML-RBF) [29] and multi-class multi-label perceptron (MMP) [23].
- Lazy and Associative Methods. In this group, Multi-Label k Nearest Neighbours (MLkNN) [27] extends the k Nearest Neighbors (kNN) lazy learning but using a bayesian approach, BRkNN [16] is an adaptation that uses BR (Binary Relevance) in conjunction with the kNN algorithm, Multi-class multi-label associative classification (MMAC) [17] is based on class association rules and CBMLC is a clustering based multi-label classification method [16]. In addition, Tsoumakas et al. propose the algorithms Random k labELsets (RAkEL) [23] and Hierarchy Of Multi-label classifiERs (HOMER) [22] to be applied in domains with large number of labels and training examples. For hierarchical multi-label classification problems (HMC) the methods HMC4.5 and Clus-HMC, HMC-LP [2] have been proposed.

3 Recommender Systems

There is a great diversity of procedures used for making recommendations in the e-commerce environment. As introduced in section 1, they can be classified into two main categories [8]: *collaborative filtering* and *content-based* approach. Nearest neighbor algorithms were originally the basis of the first class of techniques. These algorithms predict product preferences for a user based on the opinions of other users. The opinions can be obtained explicitly from the users as a rating score or by using some implicit measures from purchase records as timing logs. In the content based approach text documents are recommended by comparing between their contents and user profiles [8]. Currently there are two approaches for collaborative filtering, *memory-based* (*user-based*) and *model-based* (*item-based*) algorithms. Memory-based algorithms, also known as nearest-neighbor methods, were the earliest used [14]. They treat all user items by means of statistical techniques in order to find users with similar preferences (neighbors). The prediction of preferences (recommendation) for the active user is based on the neighborhood features. A weighted average of the product ratings of the nearest neighbors is taken for this purpose. The advantage of these algorithms is the quick incorporation of the most recent information, but they have the inconvenience that the search for neighbors in large databases is slow [15]. Model-based collaborative filtering algorithms use data mining techniques in order to develop a model of user ratings, which is used to predict user preferences.

Collaborative filtering, specially the memory-based approach, has some limitations. Rating schemes can only be applied to homogeneous domain information. Besides, sparsity and scalability are serious weaknesses which would lead to poor recommendations [3]. Sparsity is due to the number of ratings needed for prediction is greater than the number of the ratings obtained because usually collaborative filtering requires user explicit expression of personal preferences for products. The second limitation is related to performance problems in the search for neighbors in memory-based algorithms. The computer time grows linearly with both the number of customers and the number of products in the site. The lesser time required for making recommendations is an important advantage of model-based methods. This is due to the fact that the model is built off-line before the active user goes into the system, but it is applied on-line to recommend products to the active user. Therefore, time spent on building the model has no effects in the user response time since small amount of computations is required when recommendations are requested by the users. On the contrary, the memory based methods compute correlation coefficients when user is on-line. Model based methods present the drawback that recent information is not added immediately to the model but a new induction is needed in order to update the model.

The grey-sheep problem is another drawback associated with collaborative filtering methods. This problem refers to the users who have opinions that do not consistently agree or disagree with any group of users. Multiclassifiers can address this drawback by means of assignment of more than one label to the active user.

Several data mining algorithms have been applied in model based collaborative filtering; however, multi-label classification is hardly used in this application domain.

4 Multi-label Classification in Recommender Systems

Multi-label classification algorithms have been studied for the recommendation in some systems as Bibsonomy (http://www.bibsonomy.org), [20]. Quevedo [12] proposes a tag recommender system (TRS), which is based on SVM and uses logistic regression to solve the multi-label problem. In Bibsonomy every resource has multiple labels, thus, in order to help users of the social network in the labeling of the resources they handle, the system learns the labels better adapted to each specific resource and provides them to the user arranged according to the relevance that the system assigns. An algorithm based on Naive Bayes (NB) for multi-label classification of text [25] has been proposed in a visualization system for a search engine. The aim is to provide users with interesting documents among a great quantity of results.

Another example in a different application domain is the *Medical Text Indexer* (MTI) of the US National Library of Medicine (NLM), a tool that recommends more than a label for the interface used by indexers in charge of creating cites in MEDLINE, *Medical Subject Headings*(MeSH) [13] [11]. In [1] SVM method is applied in order to classify documents represented by vectors of weighted terms. In this work the Naïve Bayes classifier is also applied in order to compare the performance.

The music and the emotions it can evoke at the same time is a further research field. Multi-label classification methods such as CLR and RAkEL are useful for classifying great collections of music with multiple emotions. In that way multi-label classification has been applied in music information retrieval [18] systems and music recommender systems based on user emotions.

Most of the works in the literature are focused on labeling contents in order to recommend them later; however, multi-label classification is not applied in collaborative filtering context, which is the aim of our work.

5 Empirical Study

This study was carried out with a dataset about recommendation of books, specifically the *BookCrossing* database (http://www.bookcrossing.com). BookCrossing is a social network which aims at connecting people of the whole world across books. Since it was thrown in 2001, the web site allows in an entertaining way to trace and to share the books by users while they share experience with people from diverse levels. Readers can assess the books with a rating between 1 and 10 stars, where a value of 10 indicates a highly recommended book.

In the induction of the model that allows to predict the books of interest for the users, available demographic and geographical information about the user (age, condition and country) was used. Additionally, the reading habits of the user (attributes of the books evaluated by him: category and author) were taken into account. In the study ten classes were considered, a label for every value of rating given by the BookCrossing users. Other data needed for the study as book category were obtained from other databases as Dewey code obtained from ISBNdb.com. After some preprocessing tasks focused mainly on reducing data sparsity we made up a dataset of 5372 evaluation cases corresponding to 1222 different books, 551 authors and 203 users. Finally, the users were grouped by their demographic attributes, retaining their evaluations about the books. In this way, 3638 multi-label cases were obtained.

In the comparative study, some of the main multilabel classification algoritms were applied using the implementation available in the Mulan library (http://mulan.sourceforge.net/starting.html). The following packages were used:

mulan.classifier.transformation (methods: BR, LP), mulan.classifier.meta (methods: CBMLC, HMC, HOMER, RAkEL), mulan.classifier.lazy (methods: BRkNN, MLkNN) and mulan.classifier.neural (methods: BPMLL, MMPLearner). The empirical validation was carried out using stratified cross validation with ten folds.

Table 1 contains a summary of the results of the performed experiments for the methods that presented a better behavior. The values of the main metrics based on bipartition (Hamming Loss and Micro-averaged F-Measure, F_1), and those based on ranking (Average Precision, Coverage, OneError and Ranking Loss) are showed in the table.

Table 1 Values of some metrics for four multi-label classification algorithms

Metric	RAkEL	MLkNN	BRkNN	CBMLC
Hamming Loss	0.1262±0.0044	0.1127±0.0025	0.1131±0.0020	0.1259±0.0047
F_1	0.2017±0.0142	0.1336±0.0229	0.1660±0.0112	0.2143±0.0193
Average Precision	0.5338±0.0166	0.5839±0.0171	0.5858±0.0178	0.5358±0.0245
Coverage	2.0233±0.0782	1.7842±0.0675	1.8032±0.0679	2.0022±0.1206
OneError	0.6831±0.0264	0.6212±0.0240	0.6193±0.0274	0.6814±0.0326
Ranking Loss	0.2046±0.0098	0.1766±0.0078	0.1784±0.0089	0.2019±0.0141

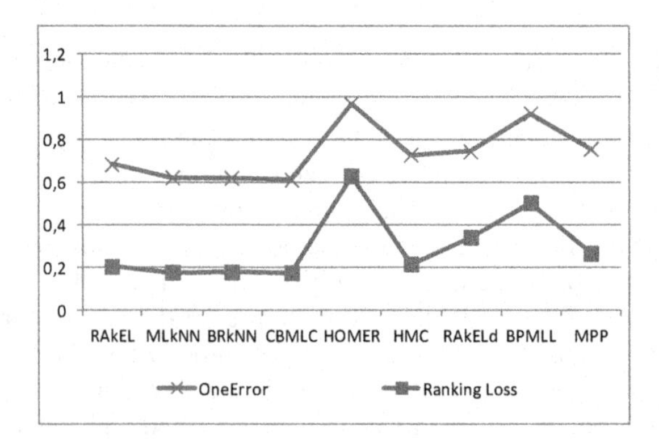

Fig. 1 Ranking metrics results

The values of Loss Hamming do not present a significant difference, which indicates the methods add not relevant labels in the same proportion. Thus, in addition the higher values of the metrics F_1 were considered. They indicated that RAkEL and CBML are better than MLkNN, BRkNN for assigning relevant labels to the instances. Besides, they present low values of Hamming loss. In addition we can deduce that CBMLC improve the behaviour of RAkEL because the increasing of F_1. For the ranking metrics these four algorithms presented similar behavior and the obtained values were better than the ones obtained with other algorithms such as HOMER or BPMLL (Figure 1).

6 Conclusions and Future Work

Multi-label classification has been hardly applied in the field of recommender systems and its application is limited to label contents for later recommendation but it has never been used in a collaborative filtering context. In this work we have demonstrated that this kind of methods can be applied for predicting user ratings about products by means of multi-label classification models that involve product and user attributes as well as ratings given by other users. Once demonstrated the utility of the main multi-label classification methods the following step will be their adaptation to this specific context and the development of a methodology for dealing with characteristic problems of recommender systems.

References

1. Aljaber, B., Martinez, D., Stokes, N., Bailey, J.: Improving mesh classificationof biomedical articles using citation contexts. Journal of Biomedical Informatics 44(5), 881–896 (2011)
2. Cerri, R., Carvalho, A.C.P.L.F.: Hierarchical multilabel classification usingtop-down label combination and artificial neural networks. In: 2010 Eleventh Brazilian Symposium on Neural Networks (SBRN), pp. 253–258 (October 2010)
3. Cho, H.C., Kim, J.K., Kim, S.H.: A Personalized Recommender System Based on Web Usage Mining and Decision Tree Induction. Expert Systems with Applications 23, 329–342 (2002)
4. Clare, A., King, R.D.: Knowledge Discovery in Multi-label Phenotype Data. In: Siebes, A., De Raedt, L. (eds.) PKDD 2001. LNCS (LNAI), vol. 2168, pp. 42–53. Springer, Heidelberg (2001)
5. Fürnkranz, J., Hüllermeier, E., Mencia, E.L., Brinker, K.: Multilabel classification via calibrated label ranking. Machine Learning 73(2), 133 (2008)
6. Ghamrawi, N., McCallum, A.: Collective multi-label classification. In: Proceedings of the 14th ACM International Conference on Information and Knowledge Management, CIKM 2005, vol. 189, p. 195. ACM Press (2005)
7. Hüllermeier, E., Fürnkranz, J., Cheng, W., Brinker, K.: Label ranking by learning pairwise preferences. Artificial Intelligence 172(16-17), 1897–1916 (2008)
8. Lee, C.H., Kim, Y.H., Rhee, P.K.: Web Personalization Expert with Combining collaborative Filtering and association Rule Mining Technique. Expert Systems with Applications 21, 131–137 (2001)
9. McCallum, A.K.: Multi-label text classification with a mixture modeltrained by em. In: AAAI 1999 Workshop on Text Learning (1999)
10. Mencia, E.L., Furnkranz, J.: Pairwise learning of multilabel classifications withperceptrons. In: IEEE International Joint Conference on Neural Networks, IJCNN 2008 (IEEE World Congress on Computational Intelligence), pp. 2899–2906 (June 2008)
11. Névéol, A., Shooshan, S.E., Humphrey, S.M., Mork, J.G., Aronson, A.R.: Arecent advance in the automatic indexing of the biomedical literature. Journal of Biomedical Informatics 42(5), 814–823 (2009)
12. Quevedo, J.R., Montañes, E., Ranilla, J., Díaz, I.: TagR: Un Sistema de Recomendación de Etiquetas basado en Regresión Logística. In: I Congreso Español de Recuperación de Información (CERI 2010), Madrid, España, de junio 15 y 16, pp. 53–63 (2010)

13. Rak, R., Kurgan, L.A., Reformat, M.: Multilabel associative classification categorization of medlineaticles into mesh keywords. IEEE Engineering in Medicine and Biology Magazine 26(2), 47–55 (2007)
14. Resnick, P., Iacovou, N., Suchack, M., Bergstrom, P., Riedl, J.: Grouplens: An open architecture for collaborative filtering of netnews. In: Proc. of ACM CSW 1994 Conference on Computer. Supported Cooperative Work, pp. 175–186 (1994)
15. Schafer, J.B., Konstant, J.A., Riedl, J.: E-Commerce Recommendation Applications. Data Mining and Knowledge Discovery 5, 115–153 (2001)
16. Spyromitros, E., Tsoumakas, G., Vlahavas, I.P.: An Empirical Study of Lazy Multilabel Classification Algorithms. In: Darzentas, J., Vouros, G.A., Vosinakis, S., Arnellos, A. (eds.) SETN 2008. LNCS (LNAI), vol. 5138, pp. 401–406. Springer, Heidelberg (2008)
17. Thabtah, F.A., Cowling, P., Peng, Y.: MMAC: A New Multi-Class, Multi-Label Associative Classification Approach. In: Fourth IEEE InternationalConference on Data Mining, ICDM 2004, pp. 217–224 (2004)
18. Trohidis, K., Tsoumakas, G., Kalliris, G., Vlahavas, I.: Multi-label classification of music by emotion. EURASIP Journal on Audio, Speech, and Music Processing 2011(1), 4 (2011)
19. Tsoumakas, G., Vlahavas, I.P.: Random *k*-labelsets: An ensemble method for multilabel classification. In: Kok, J.N., Koronacki, J., Lopez de Mantaras, R., Matwin, S., Mladenič, D., Skowron, A. (eds.) ECML 2007. LNCS (LNAI), vol. 4701, pp. 406–417. Springer, Heidelberg (2007)
20. Tsoumakas, G., Katakis, I., Vlahavas, I.: Effective and efficient multilabel classification in domains with large number of labels. In: ECML PKDD 2008 Workshop on Mining Multidimensional Data MMD 2008, pp. 30–44 (2008)
21. Tsoumakas, G., Katakis, I., Vlahavas, I.: Mining multi-label data. In: Data Mining and Knowledge Discovery Handbook, pp. 667–685. Springer US (2010)
22. Tsoumakas, G., Katakis, I., Vlahavas, I.: Multilabel text classification for automatedtag suggestion. In: Proceedings of the ECML/PKDD 2008 Workshop on Discovery Challenge, vol. 9, pp. 1–9 (2008)
23. Tsoumakas, G., Katakis, I., Vlahavas, I.: Random k-labelsetsfor multilabel classification. IEEE Transactions on Knowledge and Data Engineering 23(7), 1079–1089 (2011)
24. Ueda, N., Saito, K.: Parametric mixture models for multi-labeled text. Text 15, 737–744 (2003)
25. Wei, Z., Miao, D., Zhao, R., Xie, C., Zhang, Z.: Visualizing search results basedon multi-label classification. In: IEEE International Conference on Progress in Informatics and Computing (PIC), vol. 1, pp. 203–207 (December 2010)
26. Wieczorkowska, A., Synak, P., Ras, Z.W.: Multi-label classification of emotions in music. In: Klopotek, M.A., Wierzchon, S.T., Trojanowski, K. (eds.) IIPWM 2006. ASC, vol. 35, pp. 307–315. Springer, Heidelberg (2006)
27. Zhang, M.L., Zhou, Z.H.: ML-KNN: A lazy learning approach to multi label learning. Pattern Recognition 40(7), 2038–2048 (2007)
28. Zhang, M.L., Zhou, Z.H.: Multilabel Neural Networks with Applicationsto Functional Genomics and Text Categorization. IEEE Transactions on Knowledge and Data Engineering 18(10), 1338–1351 (2006)
29. Zhang, M.L.: ML-RBF: RRBF Neural Networks for Multi-Label Learning. Neural Processing Letters 29, 61–74 (2009)

If It's on Web It's Yours!

Abdul Mateen Rajput

Abstract. The large amount of information available on web is difficult for machine processing until or unless it is readily available in certain forms. To make it convenient for different processes and in standard pattern, it is often required that the information is available locally and as a dataset. On the web it's mostly for human use and scattered in different forms and on different locations. In this paper, we have described a method by which one can easily transform scattered information into large datasets and able to process it with different tools to get more meaningful insights.

1 Introduction

Text mining is an emerging field and there are many applications of this field since the rate of information production has increased many folds in recent past. Despite exponentially rate of data production we are still struggling for the answer of the question which can satisfy our needs as it has been said that we are drowning in sea of data while dying of thirst for knowledge. One important area which seeks answer from massive datasets is biomedical sciences, where text mining facilitates to add value and provides different procedures to analyze bulk data being produced either after each new experiment of microarray, fMRI etc or by scientific publications.

To explore the knowledge from data one needs to have access to it to get valuable information [datasets may vary in size and it depends upon the questions you are going to ask from it]. The availability of some datasets is usually restricted to the provider and user may sometime doesn't find the correct dataset he/she is interested in, though it may be browsable on the web but not available as repository to apply natural language processing and text mining tools and user finds

Abdul Mateen Rajput
Life Science Informatics,
Bonn University, Bonn, Germany
e-mail: mateenraj@gmail.com

J.B. Pérez et al. (Eds.): *Trends in Prac. Appl. of Agents & Multiagent Syst.*, AISC 221, pp. 189–192.
DOI: 10.1007/978-3-319-00563-8_23

difficulties to achieve what is required. There are many web crawlers (HTTrack[1], GRUB[2] etc) but the problem with these programs is they bring too much noise and uncleaned data. The cleaning of this data is also an issue and usually takes more time than downloading. In the current paper we discuss a smart approach to make clean dataset from any online website. The resultant dataset could be any file format you are interested in and the method will provide you different possibilities to extract from many layers of web pages. The methodology we are going to discuss is freely available and following programs are required for it:

- Mozilla Firefox[1]
- DownThemALL, Firefox Plugin[2]
- Notepad++[3]
- Linkgopher, Firefox plugin[4] /GREP (shareware) [5]

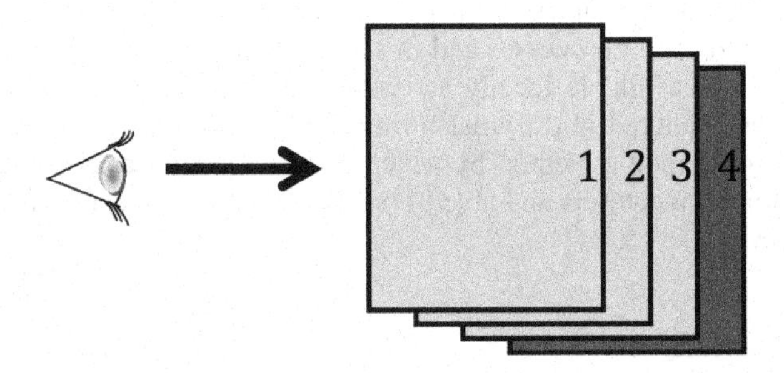

Fig. 1 A perspective from user interface. The actual dataset was below 3 web pages after searching the term. The dataset was scattered and linked beneath many different web pages.

2 Methods

The initial steps of the corpora creation requires to look for the pattern of the hyperlinks of the data you are interested in and if the links of data is available on one page then DownThemALL can automatically detects the links and you can start downloading instantly. If the actual data is under few layers of web pages then you can download the source pages and then actual data by combining all the source html pages and extracting links via LinkGopher or by using Grep program. The good feature of Grep is that it will also bring the data within the proximity of upto 5 lines from the actual search term.

In presented scenario the dataset was under many different pages and the links of the actual files were scattered on different pages. As mentioned above, we used DownThemALL to collect the top pages (all the three layers) and different term associated with the same disease (Fig. 2). The first layer contains the name of disease with different order/synonyms. There were 11 pages on second layer

[1] `http://www.httrack.com/`
[2] `http://www.gnu.org/software/grub/`

which refer to 175 pages (third layer) of relevant data files but in html format. The third layer also contains the link of actual data file (in RDF format) as hyperlink. We gathered all the 175 pages and observe the pattern of links linked to the actual data files. The manual work would have taken too long for this simple task as one has to click forward and backward thousands time and spend quit some time. With the help of tools mentioned above we were able to perform this task in less than an hour and without extra effort.

3 Use Case

The use case discusses the task we did with linkedCT.org [6], which is a RDF processed repository of clinicaltrials.gov [7]. We needed to download all the clinical trials associated with a particular disease and those clinical trials were stored under 4 different names (Multiple sclerosis Relapsing-Remitting, Relapsing-remitting Multiple Sclerosis, Relapse-Remitting Multiple Sclerosis, Relapsing Remitting Multiple Sclerosis). The actual data we were looking for was stored under 3 html pages where all the label of clinical trials associated with the disease state was mentioned (see figure 1). We stored the source html pages of actual clinical trials (4 pages associated with the disease titles) and then merge them together so we can have all the names of files on one html page. We found that the pattern of RDF storage and the page where it contains the link of it doesn't differ much and there is a similar pattern for each RDF file associated with the webpage link. Further we extracted all the links by using LinkGopher from the merged page and then looked at the patterns of RDF and html page. After finding out the pattern we simply replace the keywords with the one which was associated with RDF and then downloaded all the RDF files by simply using DownThemALL.

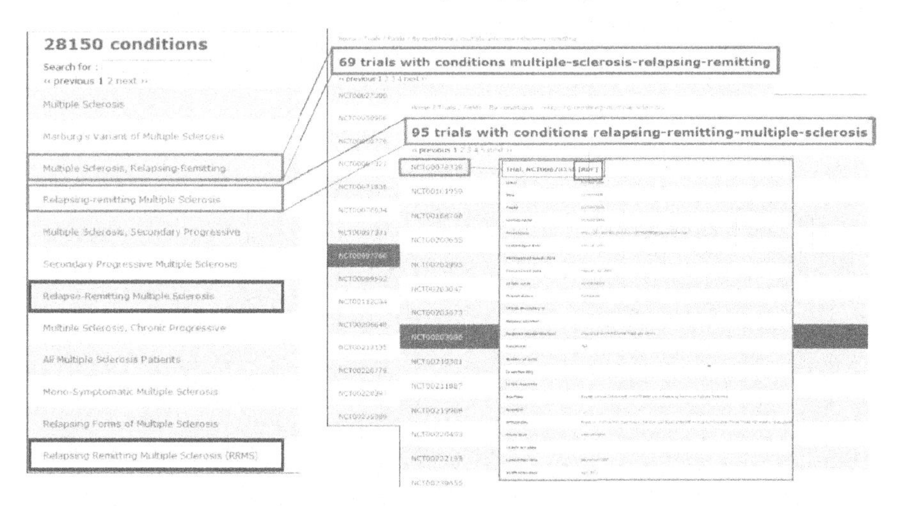

Fig. 2 The overall view of the dataset. We needed many different RDFs (in green box) stored under different pages, description page of clinical trial, label page of different clinical trial and on top the disease page.

4 Conclusion

We have used this method with several different websites and collect a large repository for using different text analytics tools. However, the procedure also has some limitation (doesn't work with Java links) and you have to carefully find out the patterns of dataset etc. On the contrary the good thing is that it is freely available and very quick rather than clicking the links and saving it manually.

5 Legal Liability

The method above describes only the technique to collect large amount of web data but it is obvious and in some cases mandatory to check the rights given to users by websites. It may be that the contents of the websites are only for human reading/visibility and prevent users to run automated procedures to aggregate information. The author doesn't take any responsibility if there will be consequences after misconduct by users and they are sole responsible for their actions.

References

1. Firefox, `http://www.mozilla.org/en-US/firefox/new/`
2. DownThemAll, `http://www.downthemall.net/`
3. Notepad++, `http://notepad-plus-plus.org/`
4. LinkGopher, `https://addons.mozilla.org/en-us/firefox/addon/link-gopher/`
5. Windows Grep, `http://www.wingrep.com/`
6. LinkedCT, `http://linkedct.org/`
7. ClinicalTrials, `http://www.clinicaltrials.gov/`

User Assistance Tool for a WebService ERP

Israel Carlos Rebollo Ruiz and Manuel Graña Romay

1 Introduction

Companies are increasingly complex requiring increasingly complex software management systems. For that reason, they resort to ERP systems, which are becoming wildly complex in an attempt to cover all the needs of the companies. They tend to have extensive menus with endless options to anticipate and try to satisfy all the information management situations. However, this software complexity taxes the human resources in the company. Complex ERP require that employees of these companies receive intensive training, and extensive further support is required of the ERP tool manufacturer.

Another problem facing companies is the increasingly common use of ERP systems over the Internet, allowing access to ERP's programs to many more users. These newcoming users often lack the necessary training to use efficiently the software resources, producing ERP failures, delays or even system shutdown, interrupting other users to receive support.

To solve this problem, Informática 68 Investigación y Desarrollo S.L. is developing a user assistance tool, aka recommender system, so that the ERP can guide users through the various programs that comprise it. Thus, we obtain two advantages. First, we reduce training requirements because the system provides a process of on-line training helping the user to know better the various programs available to him. Nevertheless, the user must have sufficient training to know what to do with the programs, because the recommender makes suggestions about the next program choice, but does not provide guides on how to use the programs. Second, we

Israel Rebollo Ruiz
Informática 68 Investigación y Desarrollo S.L.,
Computational Intelligence Group, University of the Basque Country
e-mail: beca98@gmail.com

Manuel Graña Romay
Computational Intelligence Group, University of the Basque Country
e-mail: ccpgrrom@gmail.com

J.B. Pérez et al. (Eds.): *Trends in Prac. Appl. of Agents & Multiagent Syst.*, AISC 221, pp. 193–200.
DOI: 10.1007/978-3-319-00563-8_24

enhance productivity because the system saves time by guiding the user through the steps to follow to perform their tasks, without having to navigate through endless menus.

The user assistance tool will make recommendations based on a database that will be storing each user's behavior, called User Behavior Data Base (UBDB) so that it is possible to keep track of what a particular user do at every moment. The recommender system tries to learn from the user's daily behavior taking into account the moment of the day, the day of the week, or month, because users tend to perform the same tasks in different but regularly spaced moments.

However, users can be misleading. If a user does not make a proper use of the ERP, it may cause the recommender to suggest unsuitable options for performing the required tasks. For this reason, we will create a Social Network of application users, so that some users become guides of others performing the tasks. This Social Network is not going to take into account grouping based on roles assigned by the system administrator. Using the USDB we are going to build a behavior based Social Network. Through this Social Network we can get a more accurate information using the user programs and users performing similar tasks. An additional benefit is obtained, users with bad habits redirect become more aligned with best practices in the company.

Recommender Systems

In recommender systems the Collaborative-Filtering (CF) and Content-Based Filtering (CBF) are the most used approaches. The CF [16, 2] is a method of processing the information or patterns using techniques involving collaboration among multiple agents, user points of view and any other information. The method is very usefull with big databases allowing to give good recomendations [11, 15]. The CBF method focuses more in the items than in the users [12, 9]. This method try to offer recommendation based on similar item used, or buy by the user, instead of user features.

There is no established methodology to test a recommender system. We need a way to verify the addecuation of a recomender measuring the reliability of the recommender [10] or using a especial framework [1]. Trust is a key requirement for recommender systems [32, 28].

A user's profile can be modified to fool the recommendation system to give advice concerned. We can see the problem of a user Malicious Rating Profiles recommendings in[5, 26]. This problem is not relevant for our system, because user profiles are automatically extracted from the user behavior while performing its job, instead of the user specifications.

There are a lot of fields where recomender systems can be used, like in electronic marketplace [4], in health [14], in the movie industrie [3], Web pages [7] and in new devices, such as mobile phones [30].

The rest of the paper is organized as follow: Section 2 presents ERP through the web-service. In Section 3 we discuss the user behavior data base. Section 4 introduces the previously generated Social Network. Finally, section 5 present the user assistance tool or recommender. The last section is for conclusions.

2 ERP

Enterprise Resource Planning (ERP) systems are software packages that assist an organization to manage information across finance/accounting, manufacturing, sales and service, customer relationship management, and any other task in the organization needing information processing. Usually, ERPs work on a computer or group of computers integrated in a company's network. But nowadays, with the importance that is acquiring internet, the ERP software tend to move towards the net offering agents and services accessible all over the world.

The ERP integrates all company facets, including real-time planning, manufacturing, sales, and marketing. These processes produce large amounts of enterprise data that are, in turn, used by managers and employees to handle all sorts of business tasks such as inventory control, order tracking, customer service, financing and managing human resources [24, 25].

The user can connect to the ERP through an internet access. The user exchanges information with the internet ERP's user interface, that get the information needed from databases hidden in the cloud. The system has to provide all the security firewalls to keep safe the data and the identity of the user. All the access log of each user is kept in a special database, the User Behavior Database (UBDB), storing the time when the program appear in front of the user and the time when the user close the application. The ERP user interface must be light to allow the system to provide required data in proper time. The internal task run "in the cloud" so the ERP system will need big computational and communication resources to keep the safety of the system and the integrity of the data.

We have implement the recommender system over a commercial ERP that has more than ten years in the market. The test has been made at one costumer company the has granted us access to it's UBDB in order to built the Social Network and the recommendation application. The costumer has reviewed the result to evaluate them. The ERP details and the costumer's identity are confidential information.

3 User Behavior Data Base

The ERP keeps a trace of the system's usage by the users. The system stores the program call by the user, the date and time of use. This information is stored in a real time updated database. Thanks to this database we can create the Social Network of ERP users, on the basis of the usage patterns of the diverse ERP programs, filling roles similar to the ones assigned by the system's administrator [20]. This Social Network may be time varying, according to the new data in the database.

Information stored in the UBDB is provided by the system, but not all of it will be useful to build the Social Network or sustain the recommender system. Sometimes a user opens and closes a program in so short period of time that is doubtful that he has performed any valid action with this program. Other times, a program may be open the whole day, so that it is unclear if the user was working the whole day with it, or simply forgot to close it. For these reasons, a preprocessing of the information in the UBDB must be carried out, discarding records which are not useful, or being cautious with suspecting records.

The recommender system will use all available information in the UBDB to know the programs that a user will activate by himself or induced by another user. It must identify if the program is the first one called in the day, extracting this information from the UBDB, or if the user closes one program to call another one. Therefore, the recommender will review the historical behavior of the user after closing the current program.

The UBDB stores the programs call by the user and the date of use, time running the program, but we do not intend to record all the actions performed by the user inside the program. This has some disadvantages from the recommending point of view. For instance, if the user did not perform any action on the program, that means that the program is not relevant to the task he/she is performing. However, the recommender does not have the ability to distinguish this situation and would record this program as relevant to the task performed.

Another problem is the simultaneous use of various programs. In the UBDB, each time that a program is call, there is a log stored, as well as each time that it is closed. It is possible that several concurrently opened programs have opening time, but not closing time. In this situation. The system must be able to identify the program that is useful for the accomplishment of the task at hand, in order to compile and produce adequate recommendations. However, this information is not stored in the UBDB, because it may originate an excessive data flow to the UBDB, hindering the recommender system to act in real time. Besides, when there are several programs opened, the work flow is unclear, so that the recommender will be unable to select one option as the most adequate for the task at hand. The idea in this paper is to design a first version of the recommender system that will be progressively upgraded. This alpha version will consider the sequential use of programs in a single thread.

4 Social Network

As mentioned before, the system needs a social network of users to properly guide the user through their work. The user may have little experience so using only his/her data from the UBDB will not help to advance in the learning process. The user may have bad habits that are difficult to redirect unless you perform an audit on its work, which is not the purpose of this work, quite the contrary, we want to guide the employee to be proficient autonomously, without personal or emotional pressures. We present the Social Network extracted from the ERP's logs in the paper [23] where we explain how we built the social network and how we create the group of users.

In the ERP, there are methods to aggregate users into groups such as role, department or location. But these groups are defined by the system's administrator, whom does not always know the exact tasks that each user performs. Also, if a user changes role, department or location, is very easy to forget to update this information. This causes that many users are identified in a way in the system while performing another entirely different task.

Therefore, the system creates a Social Network implicit using information from the UBDB and as shown in [31]. The Social Network will be implicit since there is not required of users to register or record in such a network. The Social Network generates relevant connections between users automatically [33, 18]. Moreover, this Social Network will be used exclusively by the recommender system being completely hidden to the system administrator or the user themselves, because it is not intended to monitor users, but to help them.

In addition, the Social Network will vary over time, because the relationship between users is recalculated each time a user calls a program, with the result that some users may leave the group to join another, or go from being lonely users to become part of a group or vice versa. However, to avoid the system to slow down, this grouping is not computed in real-time, but periodically, i.e. every night. For clustering, a previously developed graph coloring algorithm based in Swarm Intelligence [8, 22] is used, although there are many alternative methods like [6].

5 User Assistant Tool

After reviewing the use of recommenders in the literature [27], we have built a recommender based on the information generated by the ERP recorded in the UBDB and user groups that also have been generated from the UBDB. The recommender is able to offer recommendations to the users helping them in their daily work by shortening the time required to run programs, improving the level of education, and reducing errors or unnecessary access to programs.

There are generic guidelines to build recommender systems [19]. In the case of the ERP we analyze its particularities. The system calculates for each user what is the first thing the user does each time he access the ERP system to make that the first proposal of the day. Also calculated transitions between programs, to see if being in a particular program, which one is the program that runs after it most frequently. The user information is combined in each group to make a recommendation based on individual experience and that of the rest of his group. The way that the group experience affects the user can be parameterized so that if an user is a novice or inexperienced, the group's influence is bigger than his own experience, and if he is an experienced user, his experience predominates over the group, although considering this, to avoid bad practices of veteran users.

The amount of information to be processed is very large so that it will not be performed in real time, but as the group of users, it will be pre-calculated at specific moments for recommender response is at all times in real time. The response time of the system significantly affects the user experience [13].

Once the recommender system is built, it is very important the way it presents the recommendations to the user [29], because if the system is annoying, users tend to not use it, losing all its advantages. We should review the typical problems of user interface design to prevent and improve the user experience [17].

Finally, the evaluation of the recommendation system is a complex and subtle process [21]. In the case of an ERP, each user must determine whether the recommendations offered by the system adapts to his needs or not, and whether the system improves productivity or otherwise is an obstacle to the smooth functioning in their jobs.

6 Conclusions and Future Work

In the competitive development of commercial ERP the inclusion of recommender systems based on the actual working profile of the user and the collective social intelligence is thought as target innovation field.

We have designed a recommendation system for an ERP based on a UBDB and a Social Network of users based on applications and behavior. The database is updated in real time, but the system does off-line analysis of the data periodically in order to respond in real time. The recommendation system receives feedback through UBDB records that are created after the decisions adopted by the user.

In field tests carried out, we have managed to evaluate with the help of a human expert the goodness of the recommendations in controlled test cases and a UBDB with few programs and a small number of records.

As future work, the system is going to be installed and tested in more organizations using the company ERP, to verify the user experience using extensive surveys. We want to implement a feedback model for recommender beyond the pure UBDB log in order to reduce the learning time of the system itself. We should also consider more realistic scenarios such as using multiple programs simultaneously that have not been considered in this paper. Finally we must design and implement an interface to show the recommendations to the user following guidelines found in the literature.

Acknowledgements. The Informática 68 Investigación y Desarrollo S.L. company encourages participation in academic meetings, and academic production, within the limits of business confidentiality. The work reported here has been supported by project Gaitek funded by the Basque Government.

References

1. Bobadilla, J., Hernando, A., Ortega, F., Bernal, J.: A framework for collaborative filtering recommender systems. Expert Systems with Applications 38(12), 14609–14623 (2011)
2. Bobadilla, J., Ortega, F., Hernando, A., Bernal, J.: Generalization of recommender systems: Collaborative filtering extended to groups of users and restricted to groups of items. Expert Systems with Applications 39(1), 172–186 (2012)
3. Carrer-Neto, W., Hernandez-Alcaraz, M., Valencia-Garcia, R., Garcia-Sanchez, F.: Social knowledge-based recommender system. application to the movies domain. Expert Systems with Applications 39(12), 10990–11000 (2012)
4. Christidis, K., Mentzas, G.: A topic-based recommender system for electronic marketplace platforms. Expert Systems with Applications (2013)

5. Chung, C., Hsu, P., Huang, S.: A novel approach to filter out malicious rating profiles from recommender systems. Decision Support Systems (2013)
6. Firat, A., Chatterjee, S., Yilmaz, M.: Genetic clustering of social networks using random walks. Computational Statistics & Data Analysis 51(12), 6285–6294 (2007)
7. Goksedef, M., Gondiz, S.: Combination of web page recommender systems. Expert Systems with Applications 37(4), 2911–2922 (2010)
8. Graña, M., Cases, B., Hernandez, C., D'Anjou, A.: Further results on swarms solving graph coloring. In: Taniar, D., Gervasi, O., Murgante, B., Pardede, E., Apduhan, B.O. (eds.) ICCSA 2010, Part III. LNCS, vol. 6018, pp. 541–551. Springer, Heidelberg (2010)
9. Hernandez-del Olmo, F., Gaudioso, E., Martin, E.: The task of guiding in adaptive recommender systems. Expert Systems with Applications 36(2, Pt. 1), 1972–1977 (2009)
10. Hernando, A., Bobadilla, J., Ortega, F., Tejedor, J.: Incorporating reliability measurements into the predictions of a recommender system. Information Sciences 218, 1–16 (2013)
11. Huete, J., Fernandez-Luna, J.M., de Campos, J.M., Rueda-Morales, M.: Using past-prediction accuracy in recommender systems. Information Sciences 199, 78–92 (2012)
12. Kim, H., Lee, J., Wook Ahn, C.: A recommender system based on interactive evolutionary computation with data grouping. Procedia Computer Science 3, 611–616 (2011)
13. Knijnenburg, B., Willemsen, M., Gantner, Z., Soncu, H., Newell, C.: Explaining the user experience of recommender systems. User Modeling and User-Adapted Interaction 22, 441–504 (2012)
14. Lopez-Nores, M., Blanco-Fernandez, Y., Pazos-Arias, J., Gil-Solla, A.: Property-based collaborative filtering for health-aware recommender systems. Expert Systems with Applications 39(8), 7451–7457 (2012)
15. Luo, X., Xia, Y., Zhu, Q.: Incremental collaborative filtering recommender based on regularized matrix factorization. Knowledge-Based Systems 27, 271–280 (2012)
16. Mazurowski, M.: Estimating confidence of individual rating predictions in collaborative filtering recommender systems. Expert Systems with Applications (2013)
17. Nielsen, J.: The most hated advertising techniques, Jakob Nielsen Alertbox (December 6, 2004)
18. Opsahl, T., Panzarasa, P.: Clustering in weighted networks. Social Networks 31(2), 155–163 (2009)
19. Ozok, A., Fan, Q., Norcio, A.: Design guidelines for effective recommender system interfaces based on a usability criteria conceptual model: results from a college student population. Behav. Inf. Technol. 29(1), 57–83 (2010)
20. Pekec, A., Roberts, F.: The role assignment model nearly fits most social networks. Mathematical Social Sciences 41(3), 275–293 (2001)
21. Pu, P., Chen, L., Hu, R.: A user-centric evaluation framework for recommender systems. In: Proceedings of the Fifth ACM Conference on Recommender Systems, RecSys 2011, pp. 157–164. ACM, New York (2011)
22. Rebollo, I., Graña, M.: Further results of gravitational swarm intelligence for graph coloring. Nature and Biologically Inspired Computing (2011)
23. Rebollo-Ruiz, I., Graña-Romay, M.: Swarm graph coloring for the identification of user groups on erp logs. Cybernetics and Systems (2013)
24. Shapiro, J.: Bottom-up vs. top-down approaches to supply chain management and modeling. Working papers WP 4017-98., Massachusetts Institute of Technology (MIT), Sloan School of Management (1998)
25. Symeonidis, A., Chatzidimitriou, K., Kehagias, D., Mitkas, P.: An intelligent recommendation framework for erp systems. In: Hamza, M.H. (ed.) AIA 2005: Artificial Intelligence and Applications, ACTA Press, Innsbruck (2005)

26. Symeonidis, P., Nanopoulos, A., Papadopoulos, A., Manolopoulos, Y.: Collaborative recommender systems: Combining effectiveness and efficiency. Expert Systems with Applications 34(4), 2995–3013 (2008)
27. Tintarev, N., Masthoff, J.: A survey of explanations in recommender systems. In: Proceedings of the 2007 IEEE 23rd International Conference on Data Engineering Workshop, ICDEW 2007, pp. 801–810. IEEE Computer Society, Washington, DC (2007)
28. Victor, P., Cornelis, C., De Cock, M., Pinheiro-da Silva, P.: Gradual trust and distrust in recommender systems. Fuzzy Sets and Systems 160(10), 1367–1382 (2009)
29. Xu, N.: Explanation Interfaces in Recommender Systems. PhD thesis, Leiden Institute of Advanced Computer Science (2007)
30. Yang, W., Cheng, H., Dia, J.: A location-aware recommender system for mobile shopping environments. Expert Systems with Applications 34(1), 437–445 (2008)
31. Yegnanarayanan, V., UmaMaheswari, G.K.: Graph models for social relations. Electronic Notes in Discrete Mathematics 33, 101–108 (2009); International Conference on Graph Theory and its Applications
32. Yuan, W., Guan, D., Lee, Y., Lee, S., Hur, S.: Improved trust-aware recommender system using small-worldness of trust networks. Knowledge-Based Systems 23(3), 232–238 (2010)
33. Zhao, P., Zhang, C.: A new clustering method and its application in social networks. Pattern Recognition Letters 32(15), 2109–2118 (2011)

TV-SeriesRec: A Recommender System Based on Fuzzy Associative Classification and Semantic Information

Diego Sánchez-Moreno, Ana Belén Gil, and María N. Moreno

Abstract. Recommender systems have become essential in many web sites, especially in the e-commerce area; however, they are not extended enough in some domains. In this work, a recommender system for TV series is presented due to the increasing interest for this kind of products. The system implements a methodology that deals with the most important problems of recommender systems.

Keywords: Semantic Web Mining, Recommender Systems, Fuzzy Associative Classification, TV series.

1 Introduction

The growing interest for TV series that exists nowadays remains clear for the progressive increase of their offer in the broadcast programming of TV channels as well as for the high number of downloads from Internet and the birth of web sites that allow to organize and to classify them. On the other hand, the great quantity of forums arisen recently where the users request and receive recommendations from other users about TV series they might be interested in, reveals the need to develop systems for recommendation of this type of products. In spite of this demand, TV recommender systems are very scarce and the available ones either are endowed with very simple recommendation methods or the series constitute secondary products in the system.

In this work, a TV series recommender system based on semantic web mining has been developed. It is endowed with a hybrid recommendation methodology combining a clustering technique and a fuzzy associative classification algorithm

Diego Sánchez-Moreno · Ana Belén Gil · María N. Moreno
Dept. of Computing and Automatic, University of Salamanca,
Plaza de los Caídos s/n, 37008 Salamanca
e-mail: {sanchez91,abg,mmg}@usal.es

J.B. Pérez et al. (Eds.): *Trends in Prac. Appl. of Agents & Multiagent Syst.*, AISC 221, pp. 201–208.
DOI: 10.1007/978-3-319-00563-8_25

using semantic data from a specific domain ontology. The aim of this methodology is to improve the personalization and the quality of the recommendations by means of solving some of the most important limitations of current recommender systems such as sparsity, scalability, first-rater, cold-start and grey-sheep problems.

The rest of the paper is organized as follows: Section 2 includes the fundamentals about recommender systems and references to some related works. In section 3, the recommendation methodology is described. The main characteristics of TV-SeriesRec system are presented in section 4 and finally the conclusions are provided in section 5.

2 Background

Recommender systems provide users with intelligent mechanisms to search products or services that fit their preferences. The methods used for making recommendations are diverse; however, most of them present some important drawbacks. On the one hand, traditional collaborative filtering methods using nearest neighbor techniques present severe performance and scalability problems due to the high computer time required for finding the neighbors, which grows proportionally to the number of users and products in the system. Model based methods, such as data mining algorithms, do not present this drawback since the recommender model is already built when the user accesses the system; therefore, time spent in building the model has no effects in the user response time.

Another important drawback is related to the low precision caused by the sparsity of the data. Sparsity is due to the fact that the number of product evaluations (ratings) provided by the users is lesser than the number required for making recommendations.

These two problems may be minimized by means of data mining methods, however, there are other shortcomings that may occur. The first-rater (or early-rater) problem arises when it is not possible to offer recommendations about an item that was just incorporated into the system and, therefore, has few evaluations from users. Analogously, when a new user joins the system the cold-start problem takes place because there is no information about his preferences and consequently it is not possible to provide recommendations. The grey-sheep problem is another drawback associated with collaborative filtering methods. This problem refers to the users who have opinions that do not consistently agree or disagree with any group of users.

Sparsity and scalability problems can be addressed by means of reducing the dimensionality of the database used for collaborative filtering (CF) using a technique called Singular Value Decomposition (SVD) [8]. Barragáns-Martínez et al. [2] have adapted the proposal of Vozalis and Margaritis for a hybrid system combining content-based and CF approaches in the TV program recommendation domain. SVD permits increased efficiency in the calculation of the similarities for the neighborhood formation used for generating recommendations. However, despite reducing the scalability problem, nearest neighbors methods are in themselves very time consuming and they require carrying out the similarity computation on-line (at recommender time); therefore, they can never achieve the efficiency provided by model-based CF methods that are induced off-line.

Cold-start problem solution has also been the aim of recent works. Most of them focus on finding new similarity metrics for the memory-based CF approach since traditional measures such as Pearson's correlation and cosine provide poor recommendations when the available number of ratings is little, a situation that becomes critical in the cases of the cold-start and first-rater problems. In [1], a heuristic similarity measure based on the minute meanings of co-ratings is proposed in order to improve recommendation performance. Another similarity measure can be found in [3]; this is a linear combination of simple similarity measures obtained by using optimization techniques based on neural networks.

A different approach is given in [4], where a hybrid recommendation procedure is proposed. It makes use of Cross-Level Association RulEs (CLARE) to integrate content information about domain items into collaborative filters. In that way, cold-start problem can be solved by means of inducing user preferences from associations between a given item's attributes and other domain items when no recommendations for that item can be generated using CF.

Hybrid content-based and CF approaches have also been applied to deal with the first-rater problem. As a representative framework we can cite RSA (Fusion of Rough-Set and Average-category-rating), which integrates multiple contents and collaborative information to predict user preferences based on the fusion of Rough-Set and Average-category-rating [7].

Most of the works in the literature focus on dealing with one specific problem without considering the remaining ones.

3 Recommendation Methodology

The methodology implemented in the recommender system for TV series constitutes a hybrid approach that integrates different data mining algorithms and semantic web technologies in order to address the problems introduced in the previous sections. The recommendations to a specific user are made by comparing his preferences with the ones of other users but also taking into account features of users and products, which are structured according to a specific ontology.

The recommendation framework consists of two clearly differentiated parts corresponding to different processes included in the methodology. The first part corresponds to the process of construction of the recommendation models by means of data mining algorithms, whereas the second one refers to the procedure of using these models for the classification of the active user at recommendation time in order to provide him with personalized recommendations. The first part is executed off-line, before the entry of the active user in the system, whereas the second one is executed on-line, when the user requests the recommendation.

3.1 Induction of Recommendation Models

The off-line process is represented in the figure 1 by means of two activity diagrams corresponding to the two stages required for inducing the models.

The first stage involves the generation of groups of users with similar preferences and characteristics. A clustering algorithm is applied for this task by using attributes containing demographic information about users (such as age, city, profession, …) and also attributes concerning items to be recommended, which users have rated. The information about user preferences comes from the transactions they have carried out in the system. The examples provided as input to the clustering algorithm are formed by these transactions and the corresponding attributes from users and items.

The output of the clustering algorithm is a set $G = \{g_1, g_2, g_3, \ldots, g_N\}$ of users' groups, where N is a predefined number of groups which may be set according to the number of users and items available in the system. The set G is provided as input to the next step, which is responsible for assigning an ordered list of items (or products) $P = \{p_1, p_2, p_3, \ldots, p_m\}$ to each group g_i. The top items are the ones who received better evaluation from the users of the group, or the most frequent ones (taking into account the number of purchases or given ratings) or any other criterion defined by an expert in the domain area involving the system. We consider the items' frequency by means of counting the number of accesses to the items in the group. The ordered list of items assigned to the user groups will be supplied as input to the recommendation process, which constitutes the second part of the methodology.

The second stage in the construction of the recommendation models is the induction of the associative classification rules by means of the CBA-Fuzzy algorithm [5]. This technique is used to generate the fuzzy rules composing the associative classification model employed for making recommendations. The application of CBA-Fuzzy algorithm provide two important advantages: First, more reliable recommendations can be obtained since associative classification has a better behavior than other methods in sparse data contexts such as those from recommender systems [6]. On the other hand, the fuzzy rules allow the classification of the user in more than one group with different belonging degrees, dealing in this way with other important drawbacks of recommender systems, the gray sheep problem. These rules will be responsible for classifying every new user at recommender time.

As showed in the second part of the figure 1, the rule generation process has two input sets: the groups of users provided as output by the clustering algorithm and the same input data set used for building the groups. The first activity for generating the list of classification rules is to combine the two inputs. Taking into account the examples composing users' groups, each record of the training set is labeled with the identifier of the group giving as a result a new training set, which will be the input for the CBA-Fuzzy algorithm. The output provided by the algorithm is a set of classification rules $R(g_i) = \{r_1, r_2, r_3, \ldots, r_p\}$, $\forall\ g_i \in G$. Thus, the classification model is composed of a set of class association rules available for each group of users. Before running the CBA-Fuzzy algorithm, a minimum threshold value of support and confidence must be set up. It is recommended to set a high value for confidence and a low value for the support, especially in a scenario involving recommender systems, where we usually have sparse data

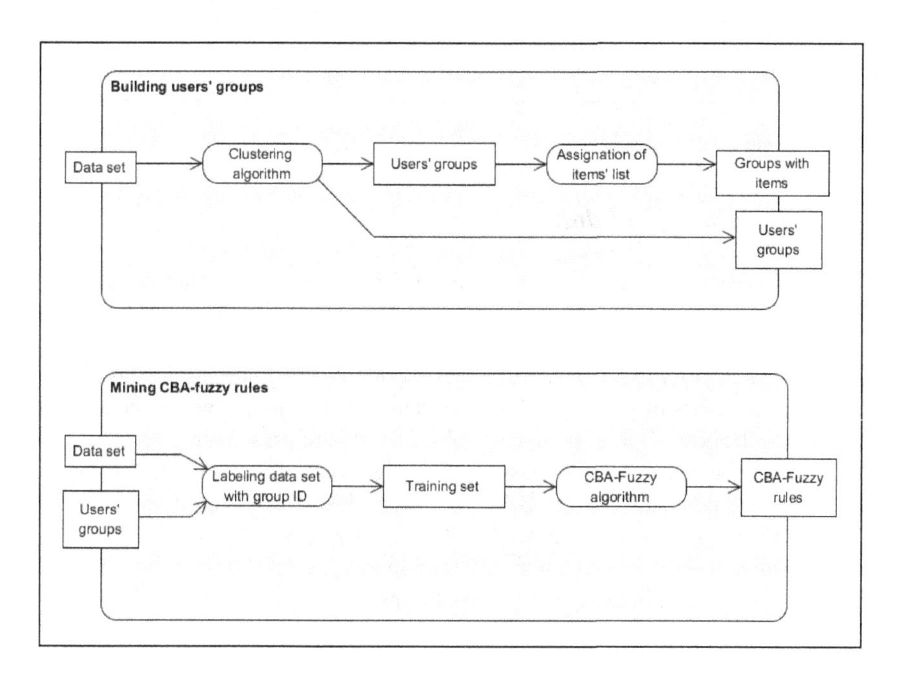

Fig. 1 Building the recommendation models

3.2 *Recommendation Process*

The models built in the previous process are used for recommending items to the active user when he is on-line. Firstly, the model of class association rules is required to classify the active user and predict in this way the group or groups he belongs to. Since preferences may change as time goes by, the most recent interaction data of the active user is taken to do the classification. In case of the active user has not done any transaction, the recommender procedure considers just the user attributes defined by a domain ontology, avoiding in this way the cold-start problem.

Figure 2 shows an activity diagram of this process, which is carried out at runtime, when the user is interacting with the system.

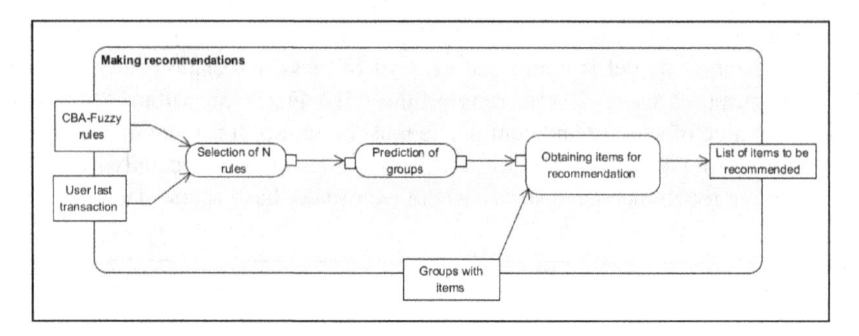

Fig. 2 Recommendation process

3.3 *CBA-Fuzzy Validation*

Given that associative classifiers present a better behavior than other classifiers in sparse data context [6], the validation of the CBA-Fuzzy algorithm has been done by comparing its precision to the results obtained with other associative classifiers (CBA, CPAR and CMAR). The study was carried out by using two datasets with very different levels of sparsity (inverse of density) that are shown in the left side graph of the figure 3. The right side graph of the figure presents the precision obtained with several associative classification algorithms for both datasets [5]. As we can see, the precision of the CBA-Fuzzy algorithm is only slightly exceeded by CMAR with the MovieLens dataset but it decreases drastically for BookCrossing dataset, which presents very higher values of sparsity. CBA-Fuzzy maintains good precision results even in high sparsity conditions.

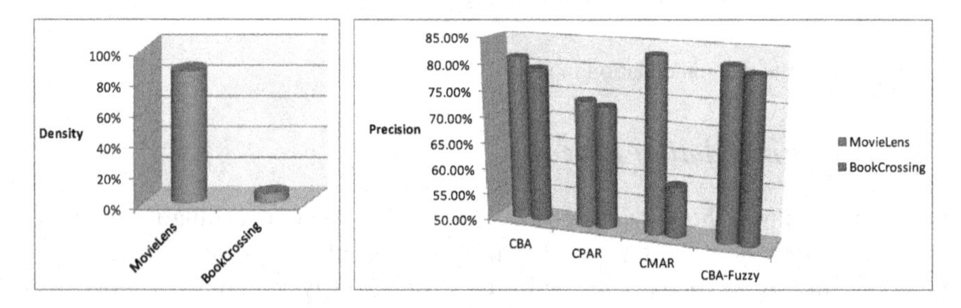

Fig. 3 Results obtained with MovieLens and BookCrossing datasets

4 TV-SeriesRec System

The methodology described in the previous section has been tailored to the TV series domain and implemented in the TV-SeriesRec system.

The recommendation methodology will allow to predict the preferences of every user on the basis of different attributes related to his personal characteristics

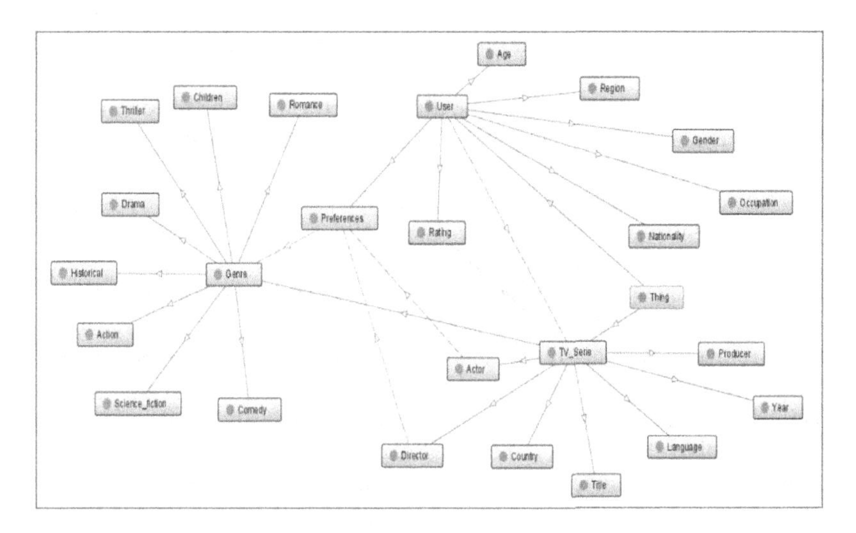

Fig. 4 Ontology for users and TV series

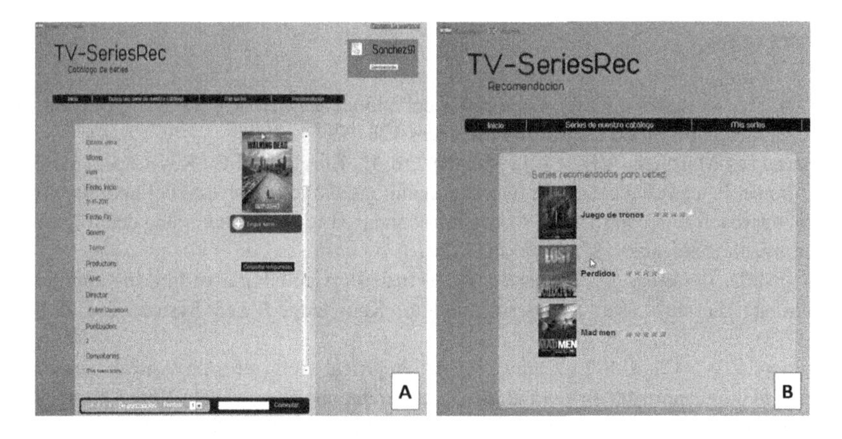

Fig. 5 Screen captures (**A.** Information about a series, **B.** Recommended series)

as well as to preferences of other users and the own characteristics of the TV series. All this information is organized according to an ontology defined specifically for this application domain (figure 4).

Apart from the recommendation functionalities, the application is endowed with catalogue management functions and provides to the users with services of search and recovery of series information according to different criteria.

Only registered users can request personalized recommendations. These users must register in order to store his personal information, which is necessary for building the recommendation models by means of the web mining techniques. Likewise these users have the possibility of evaluating the series and making comments, becoming followers of some of them, receiving notices about news, etc. Fig. 5 shows pages with information about a series and the interaction mechanisms to get some of these options (A) and recommended series to a user (B).

The system has been implemented entirely in *Java* language making use of the JSF (*Java Server Faces*) framework and MySQL as database management system.

5 Conclusions

The system presented in this work tries to supply the lack of recommender systems for TV Series that are nowadays very demanded products. The system incorporates a semantic web mining methodology for dealing with the main problems of recommender systems. The fact of being a model based approach avoids the scalability problem. The loss of precision caused by sparsity is minimized by using an associative classification method. Fuzzy rules, induced by the CBA-Fuzzy algorithm, allow to classify a user in more than a group avoiding in this way the grey-sheep problem. Finally, semantic information is used for solving first-rater and cold-start problem.

References

1. Ahn, H.J.: A new similarity measure for collaborative filtering to alleviate the new user cold-starting problem. Information Sciences 178, 37–51 (2008)
2. Barragáns-Martínez, A.B., Costa-Montenegro, E., Burguillo, J.C., Rey-López, M., Mikic-Fonte, F.A., Peleteiro, A.: A hybrid content-based and item-based collaborative filtering approach to recommend TV programs enhanced with singular value decomposition. Information Sciences 180, 4290–4311 (2010)
3. Bobadilla, J., Ortega, F., Hernando, A., Bernal, J.: A collaborative filtering approach to mitigate the new user cold start problem. Knowledge-Based Systems 26, 225–238 (2012)
4. Leung, C.W., Chan, S.C., Chung, F.: An empirical study of a cross-level association rule mining approach to cold-start recommendations. Knowledge-Based Systems 21, 515–529 (2008)
5. Lucas, J.P., Laurent, A., Moreno, M.N., Teisseire, M.: A fuzzy associative classification approach for recommender systems. International Journal of Uncertainty, Fuzziness and Knowledge-Based Systems 20(4), 579–617 (2012a)
6. Lucas, J.P., Segrera, S., Moreno, M.N.: Making use of associative classifiers in order to alleviate typical drawbacks in recommender systems. Expert Systems with Applications 39(1), 1273–1283 (2012b)
7. Su, J.H., Wang, B.W., Hsiao, C.Y., Tseng, V.S.: Personalized rough-set-based recommendation by integrating multiple contents and collaborative information. Information Sciences 180, 113–131 (2010)
8. Vozalis, M.G., Margaritis, K.G.: Applying SVD on item-based filtering. In: Proceedings of the 5th International Conference on Intelligent Systems Design and Applications (ISDA 2005), pp. 464–469 (2005)

Author Index